D0421964

TOXIC AND HAZARDOUS WASTE DISPOSAL

Volume 4
New and Promising Ultimate Disposal Options

TOXIC AND HAZARDOUS WASTE DISPOSAL

Volume 4
New and Promising Ultimate Disposal Options

edited by

ROBERT B. POJASEK
Vice President and Technical Director
Roy F. Weston, Inc.
Woburn, Massachusetts

Second Printing, 1982

230 Collingwood, P.O. Box 1425, Ann Arbor, Michigan 48106

Library of Congress Catalog Card Number 79-056116
ISBN 0-250-40265-3

Manufactured in the United States of America

Butterworths, Ltd., Borough Green, Sevenoaks
Kent TN15 8PH, England

PREFACE

I am completing this book on the occasion of my starting a new position with Roy F. Weston, Inc. This position, Technical Director of Hazardous/Toxic Substances Management, will enable me to join with many colleagues and chapter authors in attempting to solve the waste disposal problems of the 1980s. With the implementation of rules and regulations under the Resource Conservation and Recovery Act and a heightened public awareness, this is indeed an exciting and challenging time for waste management professionals.

The contributors to this volume discuss a number of the disposal options available to the generators of hazardous wastes. Unfortunately no single process can effectively handle every waste. However, any waste may often have more than one disposal option available to it. The waste management approach for each waste must objectively examine each of the applicable disposal options and in an orderly manner decide which best fits the situation. If the regulator or person performing the evaluation has a particular bias, this approach is partially void and the end disposal result may be less desirable. The reader must keep an open mind on these and other technologies which can be brought to bear in order to treat and dispose the wastes.

I would like to thank the chapter authors for their contributions. As always, I would like to thank my wife, Janice, and my sons Kevin and Andrew. Without their continued support, I would not have been able to spend the time necessary to continue this book series.

Robert B. Pojasek is Vice President and Technical Director for Roy F. Weston, Inc., Woburn, Massachusetts. He has held similar positions with Energy Resources Company, JBF Scientific Corporation, the Lawler, Matusky & Skelly engineering firm and New England Research, Inc.

Dr. Pojasek received his AB from Rutgers University and his PhD in Chemistry from the University of Massachusetts–Amherst.

An Executive Board Member of the Environmental Chemistry Division of the American Chemical Society, he is also active in the Analytical Chemistry Division of ACS and the Water Pollution Control Federation.

Dr. Pojasek is the author of over 50 publications, technical reports and presentations at scientific conferences. His current interests include design of multimedia environmental analytical programs, disposal of hazardous chemical wastes and quality of drinking water sources. He is also editor of *Drinking Water Quality Enhancement Through Source Protection,* published by Ann Arbor Science in 1977.

CONTENTS

CHAPTER 1

MINIMIZING RISKS IN LAND DISPOSAL OF HAZARDOUS WASTE

Amir A. Metry

Research and Technical Services
IU Conversion Systems, Inc.
Horsham, Pennsylvania

INTRODUCTION

Management and ultimate disposal of hazardous waste is subject to stringent programs, many triggered by enactment of the Resource Conservation and Recovery Act (RCRA) of 1976 (PL 94-580). Regulatory requirements and public pressure are critical elements in ultimate disposal of hazardous waste, requiring industry and waste management firms to guarantee short- and long-term environmental safety and public health protection. Different types of risks include threat to public health and safety from fire, explosion, toxicity, contamination of the food chain, genetic effects and carcinogenicity; environmental degradation from release of toxic and hazardous substances in air, water and land; socioeconomic impacts and legal risks due to loss of value of property liability suits.

It should be emphasized that absolute elimination of all risk associated with hazardous waste is not achievable. However, a realistic goal is to minimize the risk related to management and ultimate disposal of these wastes. Key preventive measures for minimizing the risk include:

1. waste control, reduction, reuse, pretreatment, detoxification and encapsulation;
2. siting waste disposal sites in environmentally favorable areas;
3. incorporating environmental and safety considerations in design, construction and operation of ultimate disposal facilities;
4. environmental surveillance and monitoring of water and air quality;
5. preparing contingency plans for counteracting spills, fires, explosions, contamination of air, water and/or land resources;
6. proper closure and perpetual care of completed disposal sites.

PRETREATMENT OF HAZARDOUS WASTE PRIOR TO ULTIMATE DISPOSAL

The objectives of hazardous waste treatment are the destruction, recovery for reuse and/or conversion of these substances to innocuous forms which are acceptable for uncontrolled disposal. Several unit processes may be required for the treatment of a given waste. In some cases, hazardous residues result from treatment which cannot be destroyed, reused or converted to innocuous forms, requiring controlled storage or disposal. Treatment technology can be grouped into the three categories: physical (including thermal), chemical and biological.

Selection of Treatment Process for a Given Waste Stream

In an EPA-OSW sponsored study [1] 47 unit operations and processes have been identified as potentially applicable for treatment of hazardous waste streams.

One of the great challenges is selecting an optimum treatment process for a given hazardous waste stream. No simple procedure for matching wastes and treatment processes does or could exist. Such selection is, and will continue to be, the responsibility of environmental engineers and waste management specialists.

Generalized Applicability of Treatment Processes

For wastes that are generally recognized as potentially hazardous, most conventional waste management technologies are inadequate in terms of providing sufficient health and environmental protection. Biological treatment (and associated clarification), for example, is not useful to treat toxic, flammable or other types of hazardous wastes. In fact, sanitary landfills offer inadequate protection for disposal of all but the most innocuous industrial wastes.

Table I provides a list of generalized waste streams that could be considered potentially hazardous if not treated and disposed of adequately. The options listed for each waste stream are those technologies that provide adequate protection. Although most of these processes/techniques are in use by at least one facility, a few are presently in "pilot" or "bench-scale" stages of development. The waste streams listed are generalized (generic) waste terms based on recent research and industrial profiles performed by various contractors for EPA-OSW assessment of industrial hazardous waste practices.

STABILIZATION AND MICROENCAPSULATION OF HAZARDOUS WASTE [2]

Disposal of hazardous waste by land burial can be accomplished by two containment modes: microencapsulation and macroencapsulation.

Table I. Generalized Treatment and Disposal Options

Potentially Hazardous Waste	Option for Adequate Health and Environmental Protection
Wastewater treatment sludges containing heavy metals	Dewatering, secure landfill; chemical treatment, secure landfill; chemical fixation, sanitary landfill
Wastewater treatment sludges containing fluorides	Secure landfill
Air pollution control ducts	Recycle (if possible); secure landfill
Air pollution control sludges	Dewatering, secure landfill
Halogenated solvents	Reclamation
Nonhalogenated solvents	Reclamation; if not reclamable, incineration
Solvent recovery residues	Incineration, ash to secure landfill; secure landfill
Degreasing solvents and sludges	Secure landfill
Organic chemical residues	Incineration; incineration with metal recovery; secure landfill
Oil refinery production wastes	Secure landfill
Leaded gasoline tank sludge	Evaporation, secure landfill
Nonleaded tank sludge	Secure landfill
Crude oil tank sludge	Secure landfill; chemical treatment, landfill
Inerts–contaminated waste	Chemical treatment, secure landfill
Coke plant ammonia waste	Secure landfill; ground sealing
Pickle liquors	Regeneration; chemical treatment, secure landfill
Pesticide wastes	Chemical treatment, secure landfill; storage
Slag scale and tailings	Secure landfill; ground sealing
Metal wastes (dry)	Recovery; secure landfill
Oil and heavy metal sludges	Oil recovery, secure landfill
Dye, chemical and pesticide containers	Rinse, sanitary landfill
Pigment, chemical and pesticide bags and packages	Incineration, ash to secure landfill
Stormwater silt from heavy industrial sites	Secure landfill

Completed lined landfills with suitable soil cover are an example of encapsulated wastes. Macroencapsulation is a more appropriate term because of the size involved in such placement. Another example of encapsulated waste is the sealing of waste, typically dried unstabilized material, in a thick plastic cover. Large-scale macroencapsulated hazardous wastes have the potential for environmental danger, especially in the case of hazardous wastes, should the encapsulating material fail. When microencapsulated waste is suitably compacted into a landfill site in accordance with good engineering principles such that a low-permeability monolithic material is produced, a further benefit results. In addition to the controlled-release properties resulting from microencapsulation, the landfilled waste has much less surface area exposed to water contact. Thus the total amount of leachable waste is greatly reduced even from the level made possible just by the stabilization or microencapsulation process.

IU Conversion System Process

The stabilization process offered by the IU Conversion System (IUCS), illustrated in Figure 1, is a microencapsulation process. The process stabilizes wastes into low-permeability, low-leaching, physically stable materials by means of pozzolanic reactions. Currently, contracts are held with 15 plants, representing stabilization of a total of 13.5 million ton/yr of waste materials.

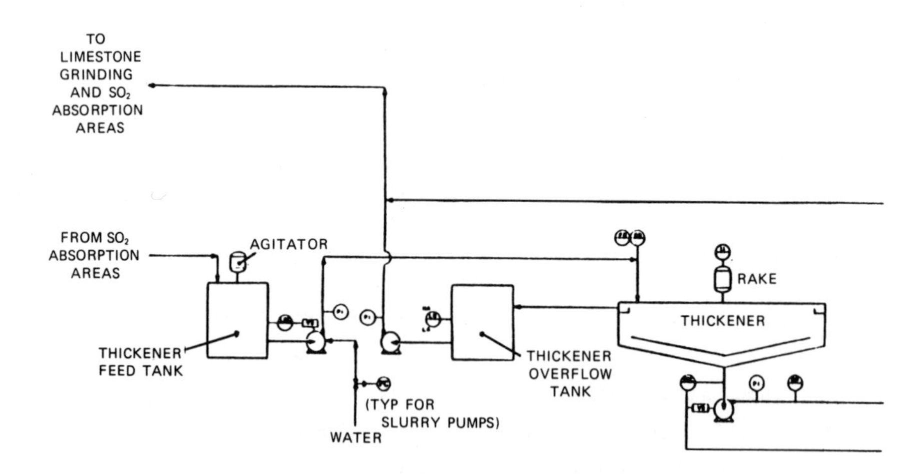

Figure 1. IUCS process for

Typically, fly ash and lime-bearing additives are used to generate these cementitious reactions. The alumina and silica in the fly ash, the lime compounds and sometimes components from the waste itself participate in several sets of simultaneous reactions. Primary reaction products include several species of calcium silicates, calcium aluminates and calcium sulfoaluminates. These compounds become hydrated in the course of reaction with as many as 32 waters of hydration attached to each molecule of unhydrated compound. Thus the process takes advantage of some of the free water typically present in waste materials, particularly sludges. In fact, it is important that the final material before placement contain sufficient water to allow these valuable reactions to proceed at their maximum rate.

Cementitious reaction products typically are a mixture of gel, semicrystalline and crystalline structures, with gels predominating. The effect is that waste particles are microencapsulated within a gel matrix. The process is in essence a microscopic version of embedding small objects within resin, except that in the case of microencapsulation some residual porosity remains, although it tends to decrease with time.

The result of the stabilization process is a cementitious material with good structural integrity and low permeability. Unconfined compressive strengths of the reacted material range from 20 to 1000 psi. Typical permeabilities are less than 5×10^{-6} cm/sec with values as low as 5×10^{-9} cm/sec having been measured. Since the reactions continue to proceed at

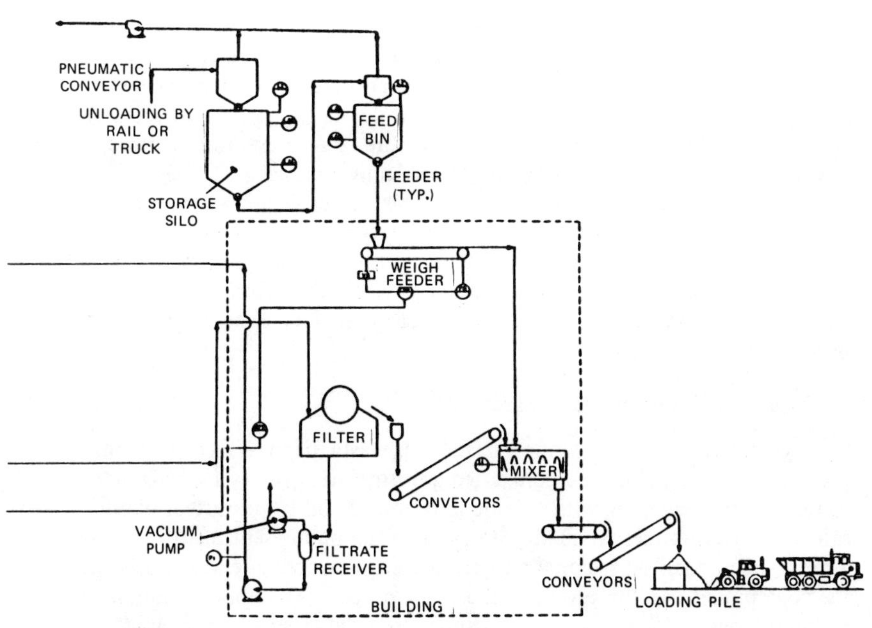

sludge stabilization.

average ambient temperature, these properties tend to improve with increasing age of the stabilized waste.

SELECTION OF ENVIRONMENTALLY ADEQUATE DISPOSAL [3]

One of the most important and complex problems associated with the establishment of a hazardous waste disposal facility is selecting a suitable location. Typically, the overall selection process involves the following basic steps: (1) developing site-selection criteria, (2) identifying candidate sites best meeting these criteria, (3) initial review and evaluation of candidate sites, (4) selection of sites for final evaluation, (5) evaluation of regional awareness, (6) final evaluation and ranking of sites, (7) public involvement, (8) site selection, (9) public hearing and (10) review.

Site selection is a complex system integrating public opinion and involvement and existing policy, while evaluating environmental, safety, economic and engineering feasibility. The relative importance of each of these factors depends on the basic selection objectives, services to be provided by the facility and pertinent local, state and federal regulations and policies.

Objectives of the Selection Process

The basic objective of hazardous waste disposal site selection is to identify potential sites that are environmentally secure, economically and technically feasible to develop, and acceptable to regulatory agencies and the public. The site must ensure the present and future safety of the public and the protection of the environment.

Site-selection criteria using environmental, socioeconomic and engineering factors must be implemented in a manner responsive to existing and new regulations, public opinion and concern, and the nature of current technology. The services to be provided and size of the facility are other factors affecting the overall objectives of the site-selection process.

Specifically, the following must be considered: projected service area (e.g., local, regional, statewide or multistate); type of treatment, such as combustion, chemical stabilization, biological treatment and land disposal; transportation access to the site; chemical and physical characteristics of the wastes accepted for disposal; planned life of the facility; long-range expansion or closure plans.

Regional and Governmental Awareness

Regional awareness for siting a hazardous waste disposal facility means presenting to the public and concerned government agencies a convincing statement of need along with a demonstration of ability to protect public safety and environmental quality. Because initial public sentiment is generally against such a facility, regional awareness means involving representatives of the government and the people in the siting process. The opinions and issues of importance to local government can aid in developing an acceptable site-selection methodology. Pressures to change the selection methodology may be experienced at the local level, with each locality suggesting another as a prime site area.

Methodology of Site Selection

Site selection of a hazardous waste disposal facility can be approached from two points of view. In either case it must be noted that final regulations have not yet been promulgated under RCRA and, therefore, complete regulations specifically related to hazardous waste disposal presently do not exist. Several states (such as California, Texas, New Jersey and Minnesota) have adopted comprehensive hazardous waste management programs; however, they may be subject to some modification as a result of RCRA.

The first approach requires that a series of site criteria must be established and evaluated with respect to their positive and negative effects; that is, the criteria must define what constitutes a good site and what does not. Major emphasis is placed on environmental adequacy and protection of public health. Basically, this approach considers the technological and physical conditions of importance for public safety and protection of the environment. Particularly for states with little or no hazardous waste regulation, the results of such an approach most likely would not fit into the framework of existing state regulations and policies specifically related to hazardous waste disposal. In this first case, the siting of a hazardous waste disposal facility is viewed as a specific problem requiring attention to specific details. The results of the siting study can be added to the existing governmental framework and can form the basis for changes to be made specifically for the establishment of chemical and hazardous waste policy.

The second approach to siting involves the examination of existing governmental regulations and policies, even though they may not specifically pertain to hazardous wastes. Facility siting is guided by the limitations and interpretations of these policies which narrow the areas of site suitability. With the exception of those states which have an existing comprehensive hazardous waste management program, a good site may be one that considers the existing political constraints first, and then evaluates the environmental adequacy and potential public health impact of the site. If the environmental and public health factors are not considered safe, then one or more of the policies must be changed.

This second approach may involve a site study conducted primarily within the political sphere, and considers the local and regional concerns of the various government entities. A site selected in this manner would be made to fit into the existing framework, allowing only case-by-case variances to be made within that framework. A determination must be made as to the nature of the existing state and local regulations and policies related to siting criteria for a hazardous waste management facility.

The first approach will be used to develop a site-selection methodology because it is not constrained by the specifics of a particular state program.

Priority objectives for site selection are stress, environmental adequacy and public health and safety.

A series of site-selection criteria are necessary to provide a base from which analysis can be completed. This allows a site or area to be evaluated with respect to specific environmental concerns as well as many of the technological aspects of the proposed project. Social, economic and political constraints can also be added to the process by using the site-selection criteria. Each set of criteria can then be evaluated to determine the positive and negative conditions with respect to siting a chemical waste facility. Essentially, an elimination process is developed by noting that certain factors are unfavorable for such a facility.

Initial analytical steps in site selection are to collect published data affecting the region proposed for the project, evaluate it and use it to limit the area of consideration. This represents an initial screening process. Sites which exhibit unsuitable characteristics can be eliminated from further consideration. Conversely, those sites with favorable siting conditions can be subjected to more detailed analysis which will lead to selection of the final site. Within this methodology, selection can begin by identifying those factors that are pertinent to the siting of a chemical waste disposal facility, selecting the factors that would constitute a favorable site and identifying those factors that would constitute an unfavorable site.

The results of these considerations can be defined as the "site-selection criteria." The physical, land-use and engineering conditions as they relate to land, water and air are included for consideration. Table II provides a summary of various factors that should be considered in the siting process and some of the favorable and unfavorable conditions associated with these factors.

PROPER MANAGEMENT AND OPERATION OF ULTIMATE DISPOSAL FACILITIES

In conjunction with criteria for hazardous waste disposal facility design, guidelines and criteria should be considered for the management of these disposal facilities. These criteria should be developed to meet federal requirements [Environmental Protection Agency (EPA), U.S. Department of Transportation (DOT), Toxic Substances Control Act (TSCA)], state and local regulatory requirements, and specific needs of the facility owner/operator for effective management. Several important aspects of facility management to be reviewed in this section include waste handling at the disposal facility, record-keeping, monitoring, financial and liability considerations.

Waste Handling

Management objectives of waste handling at the disposal facility include maintaining waste accountability, employee safety programs, management of incompatible wastes, preparing a contingency plan and proper storage.

Waste Accountability

Waste accountability must be maintained to track and record the travel of waste within the perimeter of the disposal facility and ultimately to its final disposal location. Accountability should address proper waste identification and accurate recording of waste quantities through each handling step. All hazardous waste accepted for delivery at the facility must be accounted for as receiving proper ultimate processing and disposal. This accountability, in turn, ties in with the manifest system concept which maintains waste accountability from the generator through transporter to disposal site ("cradle-to-grave" accountability).

For a secure land disposal facility accepting many types of waste, a three-dimensional grid system should be established to reflect the spatial location of all waste types. Figure 2 depicts a typical spatial grid system for landfill disposal of stacked drums. By means of this system, a particular numbered drum can be located by cell letter, grid number and grid section. Should a problem develop at the landfill, the location of all wastes and waste types is known accurately.

Employee Safety

Personnel at the facility may be exposed to hazardous wastes over long periods of time. Therefore, precautions are necessary to protect their health and safety, in addition to protecting others using the facility or visiting nearby. Facility personnel must be knowledgeable of the characteristics of the materials received. Personnel should be informed of risks to themselves and others, and should receive instruction and safety equipment for the safe handling of hazardous wastes. On-the-job safety supervision should be practiced, special supervision should be established, and emergency plans prepared. Communications should be available and emergency numbers such as fire department and medical assistance should be posted. Employees should be required to undergo physical examinations at least once a year.

The employer should provide, maintain and clean whatever safety equipment and protective clothing are required for safe work with hazardous wastes. Examples of safety equipment include impermeable suits, gloves and boots, respirators, oxygen masks, face shields, goggles, aprons, and coveralls. Specific types of protective equipment are required for some types of hazardous wastes. Decontamination equipment such as showers, chemical agents and disinfectants should be available. Special precautions should be taken during unloading and sampling of trucks to protect the site operators and truck drivers.

Incompatible Wastes

Many wastes, when mixed with others, can produce hazards through heat generation, fire, explosion or release of toxic substances, and these wastes are generally referred to as being incompatible. Waste characterization and

Table II. Sample Technical

Category	Site Characteristic	Tolerance/Suitability for
		Favorable Conditions
Land		
Soils and Topography	Topographic Relief	Gently rolling terrain
	Soils–composition, engineering and site development	Suitable soils for dike construction, building construction and liner development
	Soils–slope, erodibility	Slopes (3-10%) to limit erosion potential
	Soils–texture	Clay to silt or loam (very fine to medium grain sizes)
	Soils–agricultural uses	Soils with lesser agricultural value
	Subsoils–composition	Suitable soils for dikes, buildings and liner development
	Subsoils–permeability	Silt soils with high clay content and with low permeabilities (7-10 cm/sec or less)
	Subsoils–thickness	Thick deposits of low-permeability materials, few or no sand and gravel lens, uncompacted thickness no less than 4 ft
Geology	Bedrock–depth	Bedrock covered by thick deposits of unconsolidated material
	Bedrock–subcropping formations	Shale or undisturbed, very fine grained sedimentary formation
	Bedrock–structural conditions	No major structural variations within an area
Water	Groundwater–unconsolidated formations	No connection with surficial or buried drift aquifers, low-permeable materials to bedrock
	Groundwater–bedrock formations	Away from any recharge areas to major bedrock aquifers–no direct connection with a usable bedrock aquifer
	Groundwater–flow direction	Local flow pattern

Site Selection Criteria

Chemical Waste Disposal

Limited or Unfavorable Conditions	Considerations
Hilly, or near-steep slopes	
Poor foundation soils, unsuitable dike material; liner soils must be imported	Limited conditions will likely add to facility development costs
Slopes greater than 10% resulting in a high erosion potential	The exact slope limit needs to be defined on a site-specific basis
Fine sands to gravels (coarse grain sizes)	
Prime agricultural land	
Poor foundation conditions, unsuitable for dike materials; liner soils must be imported	Cost is an important factor in this consideration
Clean sands and gravels, with permeabilities greater than 10-15 cm/sec	Here it is assumed that natural protection of low-permeability deposits are more favorable than higher-permeability deposits
Thin deposits of low-permeable materials underlain by large thickness of sand and gravel	Ideally, a site should be underlain by a good thickness of impermeable material, underlying sands and gravels are less favorable
Bedrock at or near surface	
Highly fractured limestone or dolomites; coarse-grained, permeable sandstone	The limitations introduced by this factor are dependent on the composition and thickness of the overlying unconsolidated material
Areas of faulting, extreme fracturing or severe folding	
Underlain by surficial and buried drift aquifers of local and/or regional significance	The limitations introduced by these factors are dependent on the composition and thickness of the unconsolidated material and are site-specific
On a major bedrock aquifer recharge area; direct connection between a drift and usable bedrock aquifer	Potential for polluting a usable aquifer is the primary concern here
Regional flow pattern	

Table II,

Category	Site Characteristic	Tolerance/Suitability for Favorable Conditions
Man-Oriented	Land Use–forested	Areas where existing forests may serve as a buffer
	Land Use–cultivated land	Minor removal of land from current cultivation
	Land Use–urban residential	Areas with little urban development
	Land Use–extractive	Areas of no or low ongoing activity
	Land Use–pasture	Areas that are currently prime pasture lands
	Land Use–urban and nonresidential or mixed residential	Site-specific
	Land Use–parks, wildlife preserves, recreation areas	Very limited
	Land Use–transportation	Good-condition (9 ton or better) roads in area, lower traffic volume; near railroad
	Land Use–historical, archeological	Dependent on site-specific details
	Socioeconomic land availability	Land available for purchase; minimum amount of land owners involved
	Location	Near the majority of the waste generators
Natural Conditions	Environmental–unique areas	Area of typical regional ecosystems
	Environmental–public health	Area where construction and operation will not adversely affect public health

continued

Chemical Waste Disposal	
Limited or Unfavorable Conditions	**Considerations**
Areas where significant amounts of existing forests may be removed are not as attractive	Significant removal of forests is an additional cost factor
Areas where significant removal of prime agricultural land from cultivation is required	Significant removal of prime agricultural land from cultivation can be a local socioeconomic cost factor
Areas with high urban development	Proximity to residences is considered less favorable
Areas currently being mined or actively used	Use of abandoned extractive areas is questionable and would require site-specific investigation
Significant pasturing activities	Extent of pasturing determined by site-specific investigation
Areas with minimal commercial, industrial or institutional development	These factors can be considered exclusionary (schools, hospitals, airports)
Site location in any of these land types	All federal, state, regional, county and local parks, preserves, historical areas, etc., are considered here; this factor is considered very limited area for a chemical waste disposal facility
Roads in poor condition, high traffic volume, high-hazard roads	The absence of 9-ton roads is not exclusionary; however, upgrading of lesser roads may be a costly alternative
Areas with confirmed historical or archeological significance	This area will require some interpretation since areas of possible archeological significance have been designated; services of a professional archeologist may be required
Land unavailable or must be acquired through legal means, numerous land owners involved	
Away from waste generators	Based on the waste generator-waste disposer relationship; this aspect can become matter of transportation economics
Areas of unique ecological sensitivity, e.g., habitats of unique and/or endangered or threatened species	Extremely site-specific
Areas where dust, noise, fire, explosion may create a public health and/or safety hazard	Protection of the public is the primary consideration

Table II,

Category	Site Characteristic	Tolerance/Suitability for Favorable Conditions
Non-Development	Engineering Suitability–electric	Adequate electric power is relatively available in site area
	Engineering Suitability–sewer	Site near interceptor sewer or wastewater treatment plant
Water		
	Surface Water–waterbodies and watercourses	Limited
	Surface Water–floodplains, floodways	Limited
	Surface Water–wetlands	Limited
	Drainage–natural	Areas where surface drainage exists and can be controlled
	Drainage–local watershed	Site location near a drainage divide where upstream surface area is small
Air		
	Ambient Air Quality (odor, dust)	Good dispersive characteristics are important if the facility generates a discharge to the atmosphere
Climatology		

proper container labeling is the first step in identifying and managing incompatible wastes.

Incompatible wastes should not be mixed in the same transportation or storage container. A waste should not be added to an unwashed transportation or storage container that previously contained an incompatible waste. Incompatible wastes should not be combined in the same pond, landfill, soil-mixing area, well or burial container. An exception is the controlled neutralization of acids and alkalies in disposal areas. Containers which hold incompatible wastes should be buried well. Ideally, separate disposal areas or grids should be maintained for incompatible wastes. Incompatible wastes should not be incinerated together.

continued

Chemical Waste Disposal	
Limited or Unfavorable Conditions	Considerations
	The economics of electrical transmission are a consideration
	Not required, but could be used to dispose of clarified effluent. This item becomes a matter of economics
Placement of facility on or near	
Placement of facility on or near	
Placement of facility on or near wetlands	
Areas of poor drainage or where ponding occurs; drainage areas requiring excessive engineered controls Site location where upstream surface area is great and engineering precautions to handle runoff become costly	Site drainage may be a major factor in preventing the unplanned spread of chemical wastes at the facility; sites with suitable natural drainage conditions will be more favorable than those requiring large amounts of engineering and associated costs.
Dispersion is not expected to be an important consideration for land disposal facilities	These site characteristics are facility-specific
	Site- and facility-specific

Contingency Plan and Fire Control

Every hazardous waste treatment, storage and disposal facility should prepare a contingency plan for the protection of human health and prevention of environmental damage in the event of accidental discharge of hazardous materials. Basically, the contingency plan is an "action plan" that provides guidelines and a method for specific steps to be taken in the event of accidental or uncontrolled release of hazardous materials.

The procedures necessary to contain and recover the spilled waste and to deal with fire and explosion should be included in the contingency plan. In the event of an accidental release, the operator should act immediately to

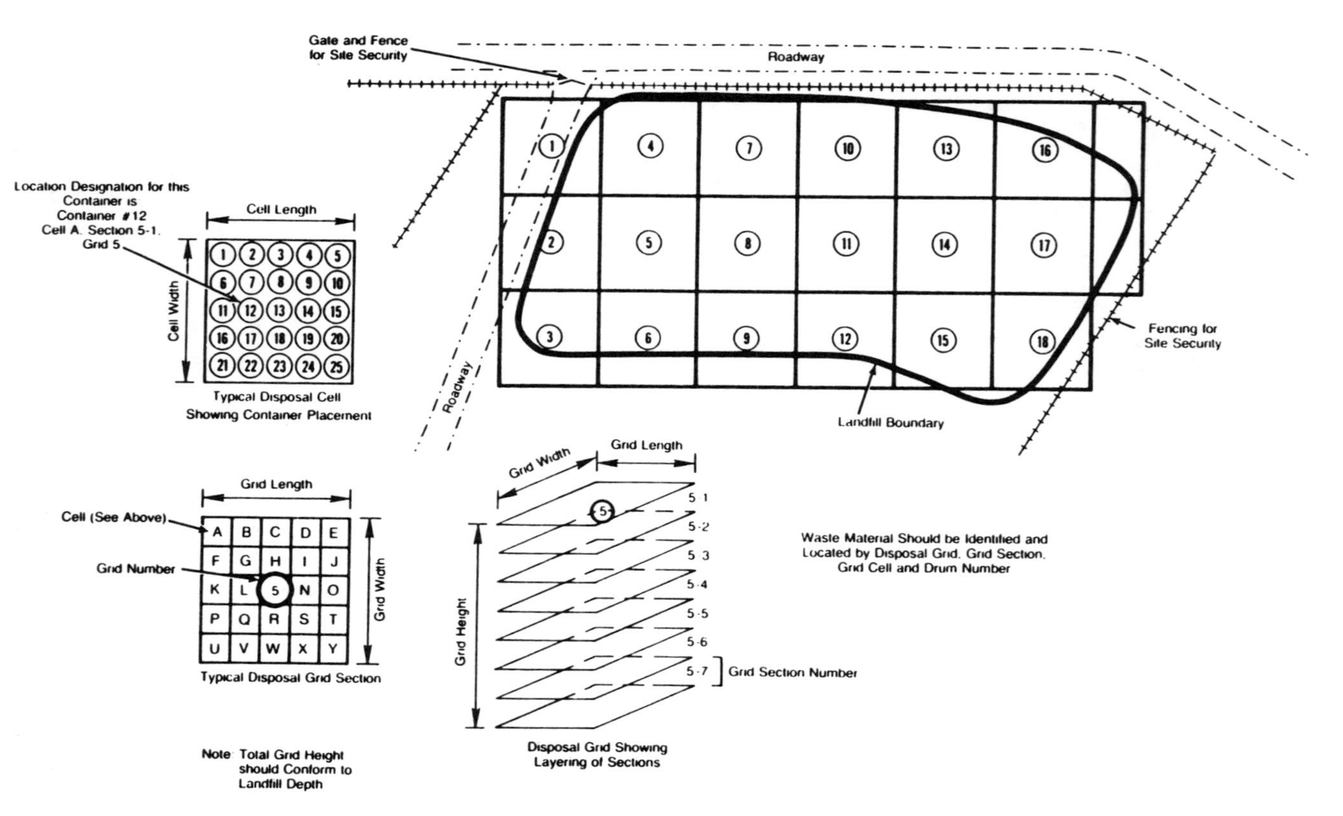

Figure 2. Typical spatial grid system for a secure landfill.

restrict and contain the movement of waste from the spill site. He should also initiate collection of the spilled waste and any contaminated material as soon as possible. It may be advantageous to deposit collected waste and contaminated material in a temporary, secure emergency storage area. From this temporary storage area, the material should be treated as a hazardous waste and transported to an approved disposal site.

Temporary Storage

Waste may often be stored temporarily at the disposal, processing or handling facility. The following is a summary of applicable guidelines.

The primary objective of hazardous waste storage operations is that they be conducted so that no accidental discharge of contaminants to the environment will occur. To support this objective, the storage units should be inspected regularly and tested for potential and actual leaks and damage repaired promptly.

Record-Keeping

Record-keeping represents an important component of the hazardous waste disposal facility management plant. In the previous section on waste handling, guidelines were developed to provide for the record-keeping associated with maintaining waste accountability. The primary record-keeping requirement for waste generators who dispose of waste at an offsite facility will involve the "manifest" documentation.

Monitoring

Monitoring is a critical tool in the management of hazardous waste facilities–directed toward maintaining proper operations and quickly detecting the uncontrolled escape of hazardous contaminants from the site.

The subject of monitoring the functioning of the facility as part of normal operations was reviewed in the sections on waste handling and record-keeping. In particular, in-plant monitoring consists of surveillance and inspection. These steps are developed to implement corrective action for equipment before failure (e.g., corrosion in a storage tank) and to correct improper operating techniques before accidents, spills or mistakes occur. Table III reflects typical items that may be included in an in-plant inspection program. Because in-plant inspections and surveillance must be developed to meet the specific requirements of individual plant operations and must be applied on a case-by-case basis, the subject will not be developed in greater depth here.

Land disposal facilities require monitoring of surface and groundwater, and heat processing or incineration requires air monitoring.

Financial and Liability Considerations

It is an important objective of the public and regulatory agencies to ensure that a hazardous waste management facility owner is financially

Table III. In-Plant Surveillance and Inspection

Storage areas for rust, corrosion, cracks in storage containers and spills
Damaged vegetation on or around the facility
Obvious air emissions and any fugitive emissions or obviously poor-quality stack emissions
General operating conditions, compliance with operating conditions as specified in the permit
Damage to fences or barriers surrounding the facility
Monitoring equipment and data
Proper materials and waste handling procedures
Dikes and drainage systems
Others as specified in permit
Inspections should be conducted daily and the results recorded in the facility's daily operating log

capable of proper management of hazardous wastes. In addition, the facility owner should be able financially to respond to unexpected accidents and damage claims. These concerns may be categorized as financial responsibility and continuity of operations.

Financial Responsibility

The disposal facility should provide financial responsibility for liability that may result from an uncontrolled, significant release of hazardous wastes. This liability may consist of damage claims, costs for cleanup and personal injury claims. Failure to provide for financial responsibility could have one or more of the following results:

- Large monetary damages could bankrupt a financially weak facility.
- Financial damages to a company-owned, onsite facility may involve the entire company.
- If the facility owner does not have sufficient financial resources, funds for performing clean-up may not be available or would be requested from federal, state or local government sources. This may result in unacceptable delays in completing clean-up.

Several alternatives for providing financial responsibility are:

- The owner/operator could post a performance bond to cover the cost of clean-up and control of an environmental occurrence.
- The owner/operator could prove, to the satisfaction of the regulatory agency, that it has sufficient assets to pay for costs and liabilities in the event of an environmental accident.
- The owner/operator could obtain adequate liability insurance for sudden, accidental occurrences in the amount related to the value of all real property

within a certain radius of the site, excluding the facility itself, to be determined by a competent appraiser and approved by the regulating agency.

- The owner/operator could maintain nonsudden, nonaccidental insurance within a minimum and maximum range to be determined by the regulatory agency. Such coverage would be for the life of the facility plus a set number of years after closure.
- The regulatory agency could encourage the local insurers to form a regional pool or to participate in national pooling of resources of those who insure such facilities for nonsudden as well as sudden occurrences.
- The state could setup a hazardous waste management fund (HWMF) that, for example, could be composed of volumetric fees on all hazardous waste entering the facility as well as fines imposed by hazardous waste legislation. The fund would be used to pay the costs of containing and cleaning up hazardous waste releases and to pay for damages to persons and property resulting from such occurrences not caused by the negligence of the owner/operator. The total amount spent for such an occurrence would be limited to a maximum amount with the state being reimbursed, if possible, by the particular HWMF.
- The federal government could go into the pollution liability insurance business. A government agency could provide first-dollar coverage or could write pollution reinsurance for private insurance companies.

The regulatory agency may either direct the facility to adopt a specific plan or allow the facility owner to select from one or more of the financial protection options and prove that he is sufficiently covered under that option. Other factors that may affect financial responsibility and the amount of financial protection necessary are as follows: (1) whether proof of financial protection will be necessary on a recurring basis; (2) whether the facility and employees will be covered under the same protection plan; (3) types and quantities of wastes; (4) type of development and land use in the vicinity of the facility; (5) natural site conditions and security; (6) total planned life of the facility and total quantity of wastes to be accepted; and (7) degree of possible impact that could be caused by an accidental release of wastes.

Continuity of Operations

Responsibility for maintaining the continuity of operations at a hazardous waste management facility may also require a financial commitment. Basically, this commitment is directed toward assuring the following operations: (1) facility closure; (2) postclosure monitoring and maintenance; (3) disposal of stored wastes; (4) maintenance of necessary disposal services to waste generators; and (5) transfer of responsibility (if necessary).

Failure to provide for continuity of operations could have one of two possible results: either a federal, state or local agency may be called upon to assume responsibility, or the facility could remain inactive without proper closure and monitoring, thereby posing a continued health and environmental threat. To provide for continuity of operations, the facility owner would probably incur an added cost such as the cost of obtaining and renewing a performance bond or payments to a sinking fund. This cost might in turn be added to the disposal rate charged to the user. In this fashion, the user (i.e., waste generator) pays both the immediate cost of disposal and the long-term facility closure care and monitoring. This concept may be viewed as the "total" or "ultimate" disposal cost.

Alternatives that have been considered as mechanisms for providing for continuity of operations are:

1. The owner/operator could prove to the regulatory agency that it has sufficient resources to close its facility and to conduct a long-term monitoring and maintenance program after closure. Annual financial reports might be used, as well as periodic engineering assessments as to the exact waste type and quantity the HWMF has been handling.
2. The owner/operator could post a performance cash bond in the name of the the regulatory agency that would, with an acceptable margin for inflation and error, pay for site closure, long-term monitoring and maintenance, and contingency plans (s).
3. The owner/operator could post a surety bond to pay for site closure, contingency plan expense and long-term site care.
4. The owner/operator could accumulate a closure and perpetual monitoring and maintenance fund by assessing a volumetric fee on all hazardous wastes entering the facility. The fund could be refunded to the owner/operator to the extent the potential problems that have been anticipated do not develop.
5. Same as Option 4 but with the fees being applied to an industry-wide trust fund.
6. Same as Option 5 but with the fund being administered by the state government or regulatory agency.
7. The owner/operator may, as an alternative to straight bonding, establish a combined bonding/fee combination. (A cash bond for closure and long-term monitoring and care deposited with an acceptable fiduciary could be withdrawn when an equivalent amount, accumulated through aggregate perpetual monitoring and maintenance fees, has been deposited by the owner/operator or a surety bond, equal to the difference between the apparent required sinking fund and the expected size of the sinking funds for that year, could be required of the site operator. The apparent sinking fund would be a site-specific reverse sufficient to provide for routine maintenance, surveillance and monitoring costs, with the option for contingency funds in the event of major facility repair.)
8. Time limitations on long-term monitoring and care. Alternatives include: (1) no limitations; (2) as part of the conditions for obtaining a permit, the owner/operator would be responsible for maintenance of its closed site for 20 years, after which it reverts to the state, or (3) government takes over the responsibility for long-term monitoring and care immediately upon closure of the facility. Financing could be obtained by a front-end fee on waste receipts.
9. Transfer of responsibility. Alternatives include: (1) the owner/operator could, as a condition to the granting of a permit, deed the site to the state; (2) the owner/operator could grant an easement which would rest in the state complete authority over the land use of the site, or (3) the owner/operator could agree to place a covenant running with the land prohibiting transfer of site maintenance with deed unless the new owner accepted and was capable of accepting the obligation. The covenant would be recorded so that a bona fide purchaser would take the land with notice.

One financial consideration is that of providing for proper closure of the facility. A basic requirement for determining the level of financial commitment is estimating the cost for closing the facility once it has reached its final capacity. This will depend on such factors as type of facility, area, size, type of wastes handled and length of time the facility is to be used.

A second financial commitment relates to assuring postclosure monitoring and maintenance. Again, the amount of financial assurance should be based on the estimated cost for this service. This will depend on such factors as type of facility, number of monitoring stations, sampling frequency, analytical tests, length of time site maintenance and monitoring is necessary, types and quantity of waste deposited in the site, and length of time the facility is to be used. Payment of premiums necessary to maintain financial protection as discussed previously should be continued after facility closure. A closed facility, as well as an operating facility, may be a cause of health and environmental damage. Costs for these continued premiums should be included as costs for postclosure facility maintenance.

In conjunction with a method for accumulating the required funds, a program should be developed for making refunds and disbursements once the need for these funds and services is no longer justified. For example, a financial commitment to assure facility closure is no longer applicable once the facility has been properly closed, inspected and approved by the regulatory agency.

CLOSING AND REHABILITATION OF ULTIMATE DISPOSAL SITES

Ultimate disposal sites should be closed after termination of operation in a manner that prevents hazards to human health and the environment. Site owners should be responsible for terminating operations and closing a site in an environmentally safe manner, and to continue site maintenance after closure.

If a hazardous waste disposal site which has closed is to be reopened for any reason, the original owner should prepare a report to subsequent owners on the maintenance requirements of the property. However, reuse of certain types of hazardous waste disposal sites should not be permitted, because of the potential hazards to users and the environment. In some instances, the state regulatory agency may find that acquiring hazardous waste disposal sites that have been closed and conducting the required surveillance and monitoring activities is environmentally safer.

Regulatory agencies may request that groundwater and surface water quality monitoring points on a site be in working condition before the closing of the facility. The owner should be requested to continue monitoring water and air quality, as appropriate, for a specific period of time after termination of operation. However, regulatory agencies should consider the necessity of long-term surveillance of all closed hazardous waste disposal sites. A monitoring and surveillance program should primarily check for possible problems (e.g., subsurface and surface water and air quality) and confirm that waste materials are not escaping from the disposal areas. Problems detected by monitoring and surveillance should be corrected immediately, and the cost of correction incurred by the original owner or by a special fund established for long-term care of such facilities.

Rehabilitation of Closed Hazardous Waste Disposal Sites

Either of two approaches for controlling pollution of subsurface or surface water by active (or abandoned) disposal sites may be taken: control of the source itself or control of receiving waters (subsurface or surface waters). In many cases, a combination of controlling the source and the receiving waters may be required.

Control strategies for abating surface and subsurface pollution by landfills can also be categorized as controls for completed, new and active (partially completed) sites.

Controls for completed sites require different technology and strategy than those required for new sites, since, in the former, waste is already in place, and in many cases subsurface pollution has already taken place. In the case of active sites, which usually contain both completed areas and areas for future use, technology for controlling existing and proposed landfills should be concurrently implemented. Figures 3 (existing landfill) and 4 (new landfill) illustrate some surface and subsurface water quality controls.

Rehabilitation of Land Disposal Site

Conventional methods for controlling leachate pollution of subsurface and surface waters (e.g., use of liners, leachate collection and treatment systems) are not expected to be feasible for most existing land disposal facilities. Experience in correcting existing sites consists basically of methods of minimizing leachate generation rates and/or interception of contaminated subsurface and surface waters. Discussion of these methods follows.

Minimizing infiltration through the site, and thus minimizing leachate generation, as shown in Figure 5, can be achieved by proper slopes of finished grades; adequate final cover and vegetation (Figure 6); diversion of runoff waters and prevention of their entry into disposal areas; construction of drainage channels and swales to speed runoff; diversion and/or blocking of subsurface water flow into disposal areas through pumping, bentonite slurry walls, etc.

Interception of groundwater inflow can be achieved by one of the techniques shown in Figure 7, which include well-point system, or a wall system; perforated drain pipe; grouting leaking areas in landfill; excavation of sand, resulting in groundwater interception and drainage through a surface conduit around the site.

Interception of contaminated runoff waters is possible at discharge points, such as springs, which show high levels of pollutants; leachate seeps from side slopes; drainage channels and swales receiving leachate; water from monitoring wells, which show high levels of pollutants.

Other Environmental Measures

Other environmental measures in closing and rehabilitating hazardous waste disposal sites should provide means for control of gaseous emissions

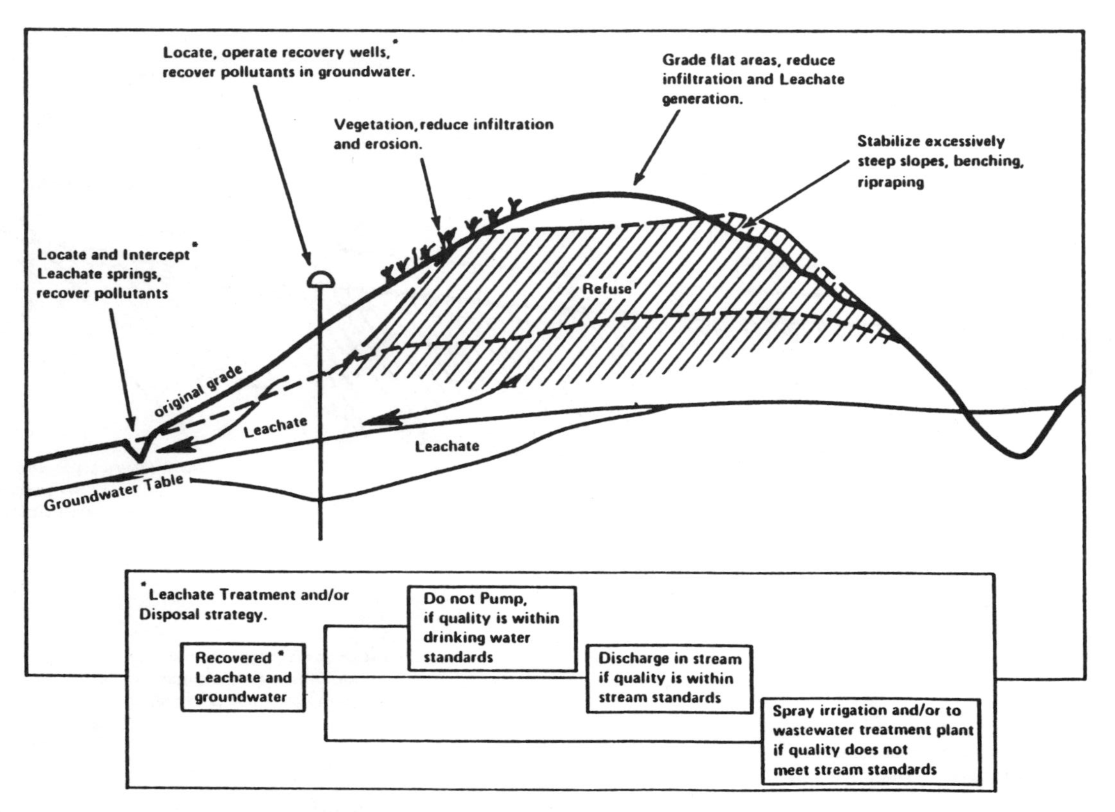

Figure 3. Means of leachate control, treatment and/or disposal for existing landfill.

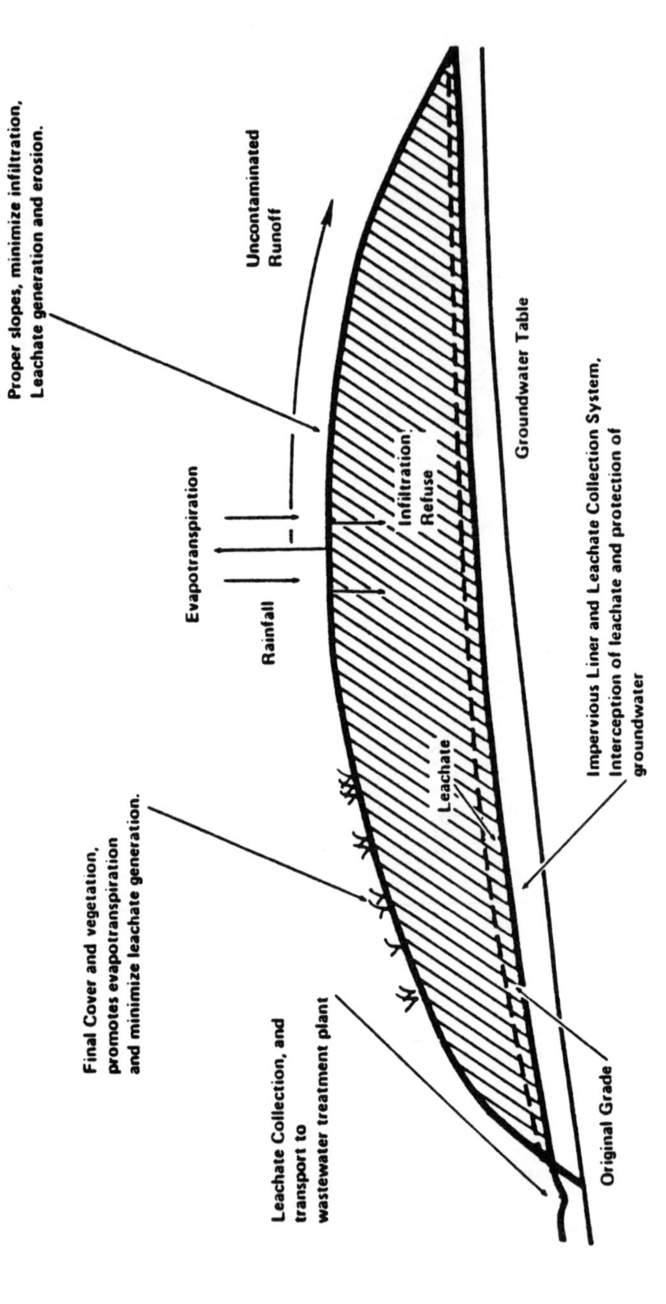

Figure 4. Means of leachate control, treatment and/or disposal for new landfill.

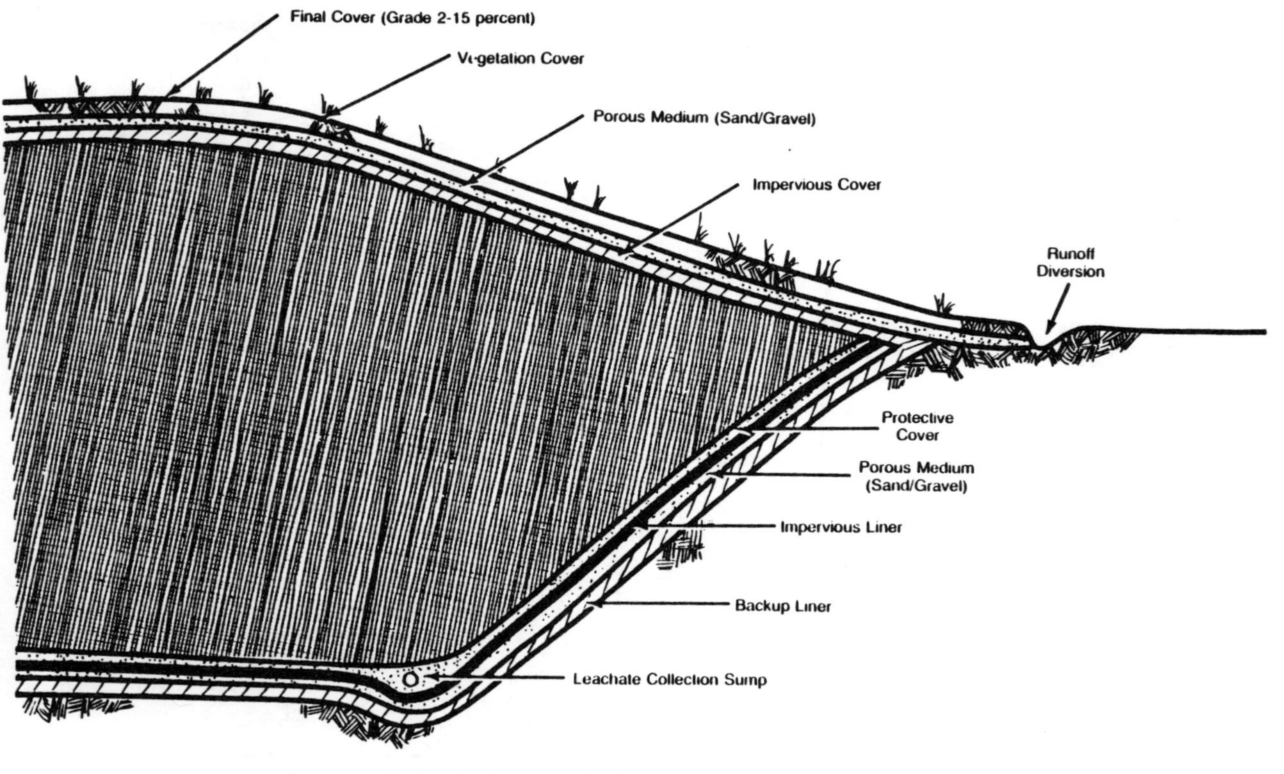

Figure 5. Means of minimizing leachate generation and waste containment.

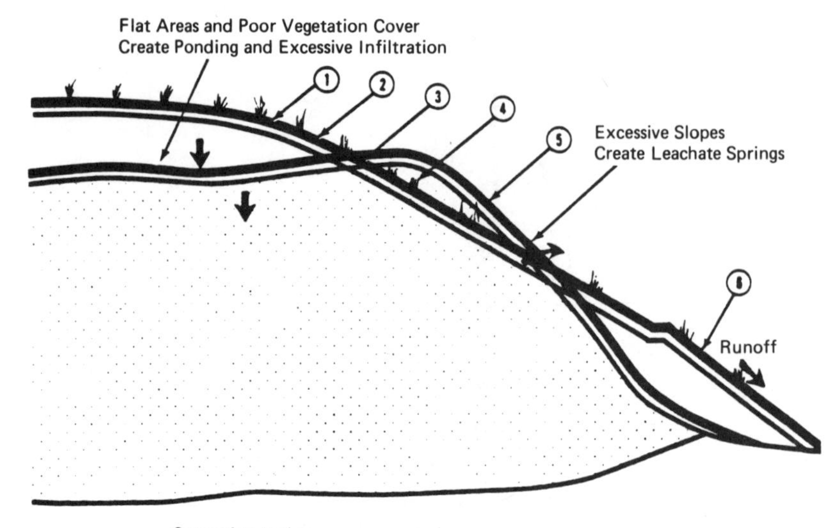

Corrective Action

1 Place an impervious cover.
2 Fill in low areas and maintain a minimum slope of 2 percent.
3 Maintain final cover at 2 ft minimum of suitable material.
4 Establish suitable vegetation cover.
5 Reduce excessive slopes to less than 15 percent or rehabilitate by benching and riprapping.
6 Provide means of quick runoff drainage.

Figure 6. Rehabilitation of landfill surfaces.

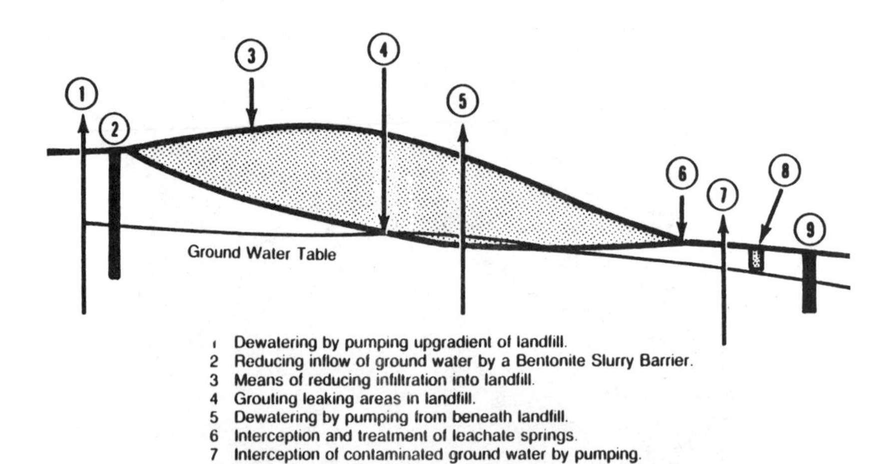

1 Dewatering by pumping upgradient of landfill.
2 Reducing inflow of ground water by a Bentonite Slurry Barrier.
3 Means of reducing infiltration into landfill.
4 Grouting leaking areas in landfill.
5 Dewatering by pumping from beneath landfill.
6 Interception and treatment of leachate springs.
7 Interception of contaminated ground water by pumping.
8 Interception of contaminated ground water by infiltration gallery.
9 Blocking contaminant flow by Bentonite Slurry Barrier.

Figure 7. Some techniques of leachate control for completed landfills.

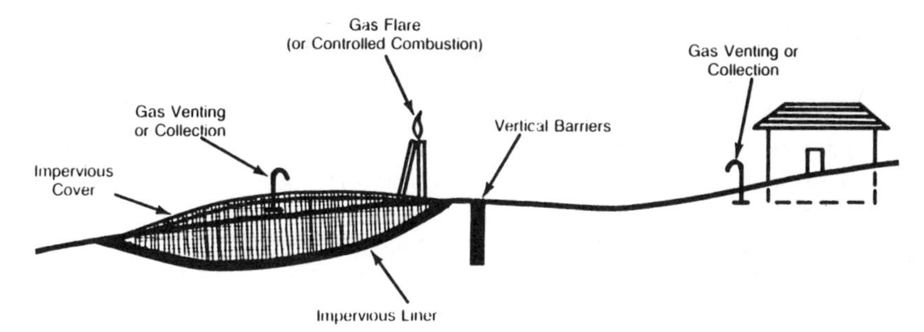

Figure 8. Means of controlling gaseous emissions in hazardous waste disposal sites.

from the disposal site. As shown in Figure 8, this should include combinations of means of preventing gases from migrating out of the disposal site, such as impervious liners or barriers; means of collecting gases, such as porous media and perforated pipes; means of venting or collecting gases; means of intercepting gases near existing structures; means of disposal or use of collected gases, such as flares, combustors or boilers.

CONCLUSIONS

New federal and state regulations subject the management of hazardous waste to stringent requirements. Ultimate disposal of such waste must be conducted to minimize risk to public health and the environment.

Pretreatment of hazardous waste, when practicable, could result in waste concentration, detoxification, recovery for reuse or conversion of hazardous elements to innocuous form before ultimate disposal.

Selection of a pretreatment process should be based on: (1) waste stream characteristics, (2) treatment requirements, (3) technical adequacy of treatment process, (4) cost effectiveness and (5) environmental acceptability.

Conventional land disposal methods do not offer safe means for ultimate disposal of hazardous waste. Microencapsulation of selected waste streams stabilizes wastes into low-permeability, low-leaching, physically stable materials. The stabilization process developed by IUCS incorporates toxic substances present in the hazardous waste in a monolithic mass by means of pozzolanic reactions.

Long-term environmental protection could be enhanced by locating ultimate disposal sites in suitable soils, hydrology, topography and land-use conditions.

Safe handling of hazardous waste at ultimate disposal sites includes: (1) maintaining waste accountability, (2) protecting employees, (3) control of incompatible waste, (4) preparation of contingency plans and (5) controlled storage and rehandling of waste.

Record-keeping represents an important component of the management and safe operation of ultimate disposal facilities. A manifest system could enhance accountability of waste throughout various handling steps.

Monitoring is a critical tool in management of ultimate disposal facilities. A complete surface and subsurface water and air monitoring program should be followed during facility operation and continued after completion of operation.

A key objective of environmental and public health protection is to ensure that a hazardous waste facility owner is financially able to respond to unexpected accidents and to provide continued monitoring and maintenance of the facility after completion of operations.

Ultimate disposal sites should be closed, after termination of operations, in a manner that prevents hazards to human health and the environment. Key elements include: control and containment of water; protection of receiving streams; rehabilitation of disposal areas; and continued site surveillance and maintenance.

REFERENCES

1. Berkowitz, J. B. et al. "Physical, Chemical and Biological Treatment Techniques for Industrial Wastes," Arthur D. Little, Inc. U.S. EPA Contract No. 68-01-3554 (November 1976).
2. Roberts, B. K., and C. L. Smith. "Microencapsulation: Simplified Hazardous Waste Disposal," in *Toxic and Hazardous Waste Disposal, Vol. 1* (Ann Arbor, MI: Ann Arbor Science Publishers, Inc., 1979).
3. Metry, A. A., M. H. Corbin and K. M. Peil. "Management Aspects of Potentially Hazardous and Special Wastes," paper presented at 3rd Annual Conference on Treatment and Disposal of Industrial Wastewaters and Residues, Houston, TX, April 1978.

CHAPTER 2

HANDLING EMERGENCIES IN HAZARDOUS WASTE MANAGEMENT FACILITIES

Enos L. Stover

Research & Development
Metcalf & Eddy, Inc.
Boston, Massachusetts

Amir A. Metry

Research & Technical Services
IU Conversion Systems, Inc.
Horsham, Pennsylvania

INTRODUCTION

A hazardous waste management system, considered from the point of generation to the point of ultimate disposition, can potentially include a number of elements, such as the following:

1. collection, concentration and storage of wastes at the point of origin;
2. processing of the waste, including deactivation, detoxification or isolation at the site of origin;
3. transport of the hazardous waste to a separate processing site;
4. processing at the separate processing site;
5. transport of hazardous residues; and
6. disposition of hazardous residues.

Proper hazardous waste management requires scientific and engineering effort before, during and after treatment and disposal. The following are basic ingredients in planning, construction and operation of hazardous waste management facilities:

1. characterization of hazardous waste;
2. study and evaluation of proposed facility site;
3. design and construction of the site;
4. pretreatment or conditioning of the wastes;

5. placement and containment of the wastes;
6. completion and closing of the facility;
7. treatment of leachate; and
8. maintenance, surveillance and environmental monitoring.

All hazardous waste should be pretreated or prepared in a proper manner to reduce potential environmental and health effects. Pretreatment and preparation procedures include the following:

- chemical fixation
- volume reduction
- waste segregation
- detoxification
- degradation
- encapsulation

Regardless of the efforts to pretreat hazardous waste, a certain amount of hazardous material will still remain. Ultimate disposal of these wastes can take place in any of the following environmental sinks:

- land–landfilling, land application, deep-well injection
- air–incineration, evaporation
- water–ocean disposal, release into surface or subsurface waters

Ultimate disposal of most hazardous wastes normally results in some form of land disposal. Landfills, as shown in Figure 1, that are designed, constructed and operated in a manner that makes them suitable for disposal of extremely hazardous wastes, are called secured, controlled, special or chemical landfills [1].

Hazardous waste handling, transport, treatment and ultimate disposal can result in many types of accidents which create emergency situations. An example of such emergency situations is the explosion in the Rollins waste disposal facility in Logan Township, NJ, in November 1977. Six people lost their lives and the total facility was ordered closed by the New Jersey Department of Environmental Protection. Under the Resource Conservation and Recovery Act (RCRA), a hazardous waste disposal system is being developed for tracking these wastes throughout their life, to ensure that they are disposed of in approved facilities that provide long-term public safety. However, these procedures will not eliminate the possibility of accidents and emergency situations at hazardous waste management facilities. The best solution for dealing with these situations is the development and implementation of accident prevention procedures and programs, as well as remedial procedures and methodologies for responding to accidents and emergency situations.

OBJECTIVES OF WASTE HANDLING TO MINIMIZE EMERGENCY SITUATIONS

Maintaining Waste Accountability

Control measures call for hazardous waste identification and classification, but classification of hazardous wastes is not an easy task. The hazardous

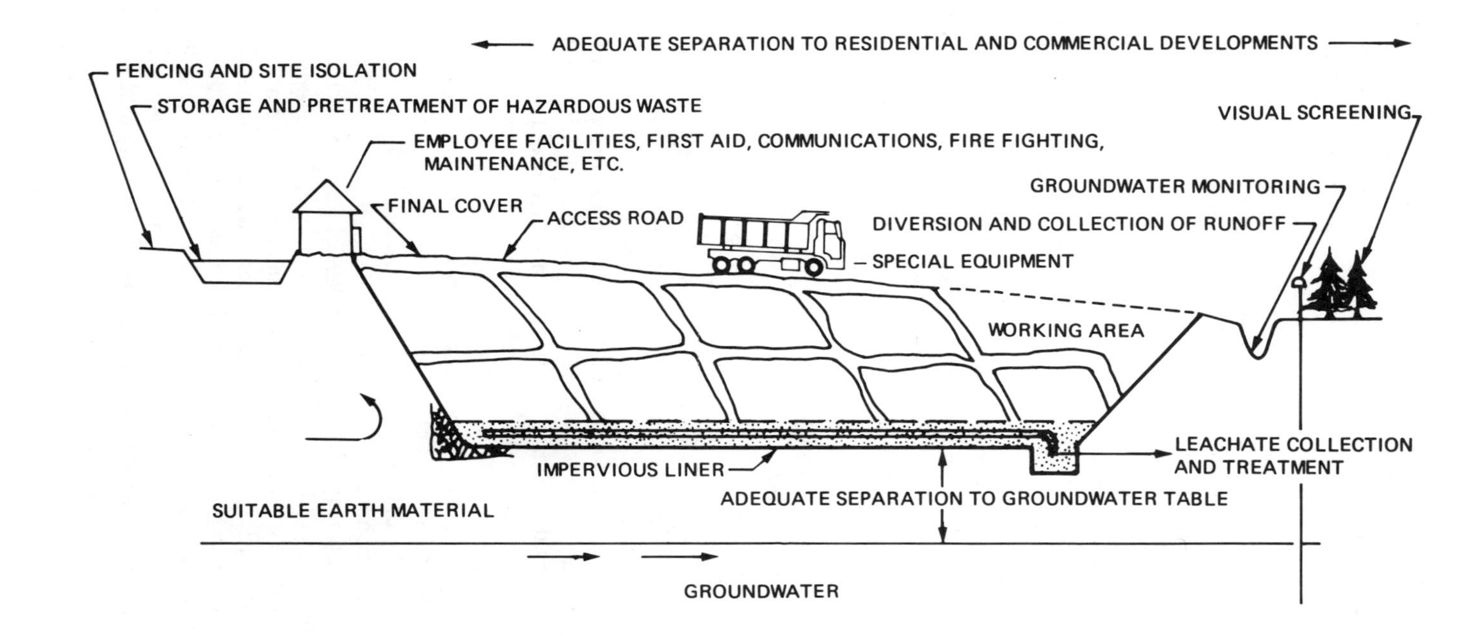

Figure 1. Elements of a controlled landfill.

material list created by the U.S. Environmental Protection Agency (EPA) contains over 250,000 entries. Hazardous waste must, however, be identified, labeled and classified, so that handlers and waste management facility operators can employ the correct technology to minimize environmental and health hazards.

All hazardous wastes quantities must be followed and tracked from "cradle to grave." Cradle to grave regulation of hazardous wastes is presently being developed by EPA under RCRA. Initial regulations propose standards applicable to transporters of hazardous wastes. EPA also plans to propose regulations applicable to generators of hazardous waste, and to owners and operators of hazardous waste storage, treatment and disposal facilities [2].

Many states have adopted or are considering adoption of a manifest system, which is primarily a permit system for transporters of hazardous waste. A manifest should be completed by the hazardous waste generator, hauler and disposal facility operator. In Figure 2, the manifest cycle is illustrated. This system is primarily concerned with tracking hazardous waste from generation to disposal. However, all hazardous waste quantities must be followed and tracked in a similar manner from delivery to the waste management facility, through storage and processing, to ultimate disposal. Proper identification, containment, storage, metering and weighing are necessary factors of the facility tracking system.

Hazardous waste must be labeled properly before storage, collection, transportation or acceptance for treatment or disposal at a waste management facility. Containers should bear the date of filling and information on the generator and transporter of the hazardous waste. These labels should be maintained on all hazardous waste containers during storage. The waste should be stored in durable, leakproof, nonabsorbent, waterproof containers such as barrels (drums).

Construction materials for containers should be dictated primarily by the corrosiveness of the waste to be handled. Polyethylene or reinforced fiberglass, either as the primary construction material or as a liner, is recommended for most types of inorganic and organic wastes except those containing hydrofluoric acid. Polyethylene-lined, 55-gal, snap-on, ring-sealed steel drums can be used for most classes of wastes since they provide strength and corrosion resistance. These containers can be used for transportation and storage at the waste management facility when they are properly identified and labeled.

Prevention, control and detection of spillage of waste at the hazardous waste management facility during storage and processing is an important concern that must be addressed. Prevention and control can best be handled primarily on the basis of selection of containers, size of containers and rate of dispersion after spillage. Detection of spillage may be accomplished by either chemical or physical means. Chemical tests employed to detect spillage may include testing the thermal conductivity of the atmosphere and spot-testing for specific waste spillage. Physical testing may include any of the following:

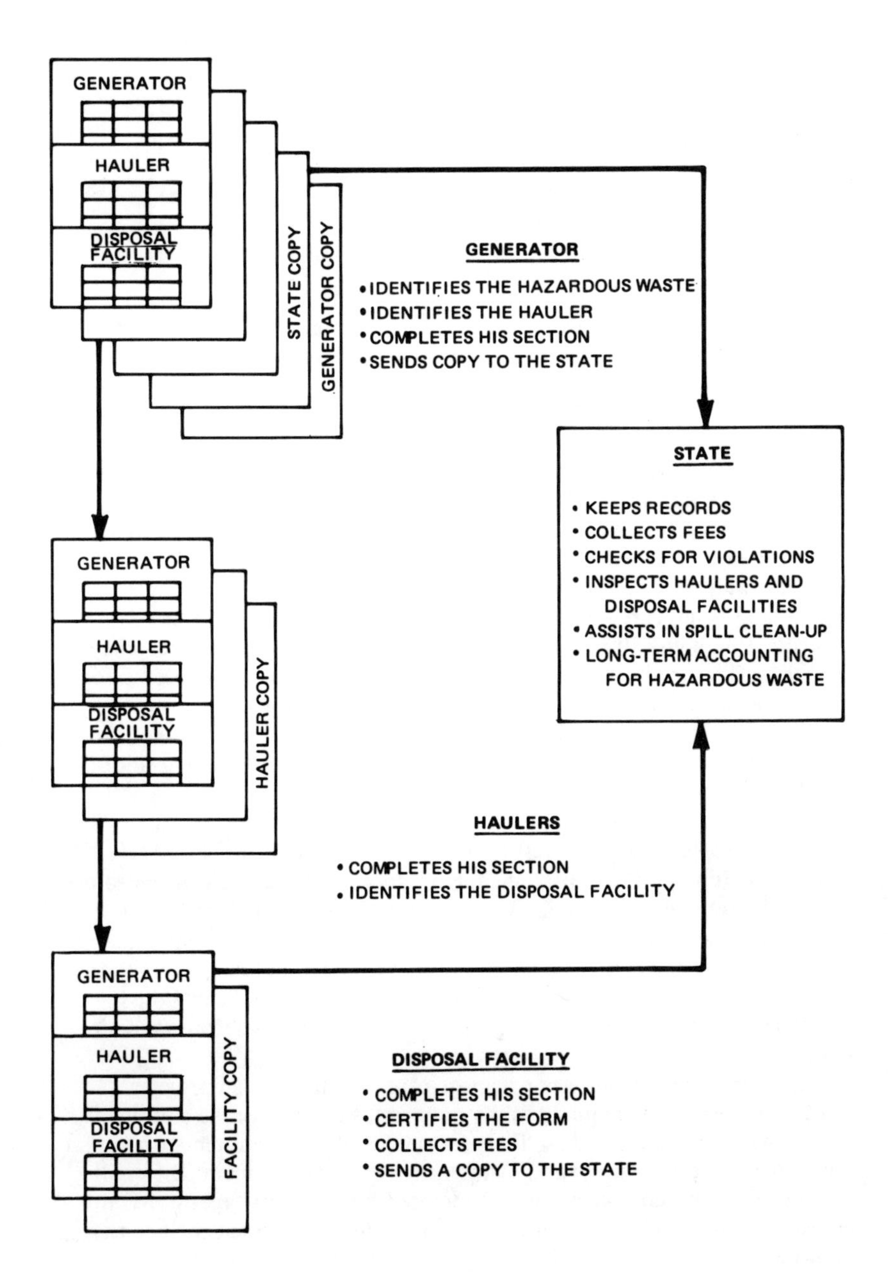

Figure 2. Manifest cycle.

1. weigh entire shipment;
2. weigh drum suspected of leaking;
3. measure liquid level in container;
4. measure conductivity of area;
5. measure light transmittance within enclosure; and/or
6. measure temperature.

Major criteria for selection of a spill detection device or system include speed of response, reliability and ruggedness.

Employee Safety

Personnel at a hazardous waste management facility may be exposed to hazardous waste for extensive periods of time. Therefore, precautions are necessary to protect their health and safety in addition to protecting others using the facility or living and visiting nearby. The personnel at the facility must be knowledgeable of the characteristics of the waste materials received for processing. They should be informed of the risks to themselves and others, and should receive instructions and safety equipment for the safe handling of the various hazardous wastes. Special supervision should be supplied for new employees, authorized visitors and regulatory personnel. On-the-job safety supervision should be practiced at all times. Accident prevention and safety programs should be established, and emergency action plans should be prepared for immediate and appropriate response to accidents. Communications should be available at all times with emergency numbers such as medical assistance and fire department numbers posted for quick response. Some of the major control programs for worker protection that should be instituted at hazardous waste management facilities include isolation, personnel protective equipment, training, personal hygiene and monitoring [3,4].

Isolation limits the potential exposure from an accident to the least number of people for the shortest time period. Isolation in space and time involves the limitation of contaminants to a specific area and allows personnel to work in that area for the minimum practical time period. Schedules and work programs can be manipulated to prevent the transport of contaminants to points outside the control areas and to prevent unauthorized access to the accident site. Typical isolation procedures may include establishing the defined contaminated area, erecting barriers and instituting decontamination procedures.

Employers at hazardous waste management facilities should provide, maintain and clean the required safety equipment and protective clothing for working with hazardous wastes. The personnel protective equipment program should include selection, use and maintenance of this equipment. Different types of protective equipment will be required for different types of hazardous wastes. Examples of personnel protective safety equipment include the following:

- impermeable gloves suitable for various tasks;
- impermeable suits, coveralls and aprons which prevent as much skin exposure as possible. This outer protective clothing may be either launderable or disposable;

- boots that are easy to clean, suitable for protection during the various operational tasks performed and impermeable to toxic hazardous wastes;
- hair and head covering such as paper or cloth caps and hard hats;
- eye protection compatible with both the toxic and mechanical hazards (goggles, face shields, etc.); and
- oxygen masks and respirators, along with appropriate respiratory protective programs.

Washing facilities, including eye wash stations, must be readily accessible from all locations throughout the facility. Special protective precautions should be taken during the unloading of trucks to protect the site operators, truck drivers and the environment.

Positive steps must be taken at the hazardous waste management facility to develop a training program in hazardous waste management for all the people working at the facility, especially the technical staff. The technology generally exists for the proper handling, treatment and disposal of hazardous waste; therefore, education and training provide the capability to deal with the problems and procedures needed for safe and efficient hazardous waste management. Personnel need to be educated in the precautions that are necessary for protection of their health and safety, as well as protection of the environment. Methods for education and training that have been used successfully include the following:

1. special training programs;
2. conferences;
3. seminars;
4. published literature;
5. libraries;
6. surveys, plans and studies;
7. EPA technical assistance programs;
8. hazardous waste advisory committee;
9. contacts at universities and industries; and
10. interaction with other facilities.

Education and training provide the means to apply control measures intelligently. Training programs should include specific instruction such as what control measures should be used, how they work and the types of hazards they can control. The program should also convey to the worker what he specifically needs to do and when.

Personal hygiene may be the single most important factor in controlling exposure of workers to hazardous wastes. Inhalation can be controlled by use of respirators, surface cover and dust control. Protective clothing and shower and shampoo before leaving the job site can control skin absorption. The shower area should include a locker area where street clothes are kept, showers and a change room where the contaminated clothing is removed. Ingestion can be controlled by removal of outer clothing and washing of face and hands before smoking and food consumption. Smoking and consumption of food and beverages should only be allowed in separate areas free from contamination after proper personal hygiene measures have been taken. Smoking and eating breaks should be scheduled for control and protection.

Workers should know basic first aid for various types of injuries and accidents. Medical consultation should be available to establish procedures for handling hazardous waste emergencies. Appropriate first aid measures should be taken immediately in the event of an accident, and persons accidentally exposed should be taken to a doctor. Plans for handling emergencies should be made in advance with a nearby physician, and a vehicle should be available onsite for transporting victims.

Incompatible Wastes

Many hazardous wastes when mixed with other wastes will produce heat, explosions, fires or release of toxic substances [5]. Producers of hazardous waste should be required to describe and characterize their wastes accurately. The type and nature of the waste, chemical composition, hazardous properties and special handling instructions should be included in the description. Table I presents some examples of types of hazardous wastes that are relevant. In cases where a hazardous waste corresponds to two or more types of waste, each relevant type must be identified. Table II is a summary of some potentially incompatible waste materials along with the potential consequences when they are mixed.

All wastes must be identified before acceptance at the hazardous waste management facility to determine the treatment and disposal scheme that should be used for each particular waste. The facility should contain an analytical laboratory with the proper equipment (atomic absorption, gas chromatography, organic carbon analysis, wet chemical analysis, etc.) to perform complete analysis of the incoming wastes. All wastes should be analyzed in the laboratory before acceptance at the facility and after arrival at the facility to confirm the composition of the waste. After chemical analysis, the type of treatment for a particular waste should be determined and the waste assigned to a specific storage area.

Unloading of hazardous waste after identification and classification must be controlled. Directions indicating where specific wastes are to be discharged from the transport vehicle must be clearly defined. Cleaning or decontamination materials should be available at the facility to wash down spilled and splashed materials. Washwater drainage or cleaning materials should be controlled and directed to proper disposal areas in the facility. Drums or tanks for holding incompatible wastes must be separated at the facility site. Storage sites must be designed, operated, located and constructed such that spills and leaks can be contained at that location within the facility.

Incompatible wastes should not be mixed in the same transportation or storage container. Wastes should not be added to unwashed containers which previously contained incompatible wastes. Incompatible wastes should not be combined in the same burial container, landfill, soil-mixing area or lagoon. Containers holding incompatible wastes should be separated by soil or refuse when they are buried. Separate disposal sites should be used for these types of wastes when possible. Incompatible wastes should not be incinerated together.

Table I. Examples of Hazardous Waste Types[a]

Type	Examples	Examples
Type 1. ACID SOLUTION		
	Spent Etching Solution	Acidic Chemical Cleaners
	Spent Acid Plating Solution	Electrolyte
	Pickling Liquor	Spent Acid
	Acid Sludge	Sulfonation Tar
	Battery Acid	Copper Bathing Solvent
Type 2. ALKALINE SOLUTION		
	Alkaline Caustic Liquids	Caustic Wastewater
	Alkaline Chemical Cleaners	Lime and Water
	Alkaline Battery Fluid	Lime Sludge
	Acetylene Sludge	Lime Wastewater
	Oakite	Lime Soda Water
	Wyandotte Cleaner	Spent Caustic
		Spent Cyanide Plating Solutions
Type 3. PESTICIDES		
	Unwanted or Waste Pesticides	Unrinsed Pesticide Containers
	Pesticide-Containing Wastes from Pesticide Production or Formation	Washwater from Cleaning Pesticide Containers or Application Equipment
Type 4. PAINT SLUDGE		
	Paint Slops	Paint Waste from Paint Production and Application
	Pigment Sludges from Paint Production	
Type 5. SOLVENT		
	Cleaning Solvents	Paint Remover or Stripper
	Data Processing Fluid	Dry Cleaning Wastes and Other Spent Cleaning Fluids
	Attrix Solvent	
Type 6. TETRAETHYL LEAD SLUDGE		
	Wastes from Tetraethyl Lead Production	Sediments Containing Tetraethyl and Other Organic Lead
Type 7. CHEMICAL TOILET WASTES		
Type 8. TANK BOTTOM SEDIMENT		
Type 9. OIL		
	Floc	Crude Petroleum
	Oil Sludge	Bleacher House Waste Oil
	Refinery Waste	
Type 10. DRILLING MUD		
Type 11. CONTAMINATED SOIL AND SAND		
	Sand and Oil	Lagoon Residue Mixed with Soil
	Spent Blasting Sand	Contaminated Soil or Sand from Spills

[a]Source: "Hazardous Waste Management–Law, Regulations and Guidelines for the Handling of Hazardous Waste," California State Department of Health (February 1975).

Table II. List of Potentially Incompatible Wastes[a]

Group 1-A[b]	Group 1-B
Acetylene sludge	Acid sludge
Alkaline caustic liquids	Acid and water
Alkaline cleaner	Battery acid
Alkaline corrosive liquids	Chemical cleaners
Alkaline corrosive battery fluid	Electrolyte, acid
Caustic wastewater	Etching acid liquid or solvent
Lime sludge and other corrosive alkalies	Liquid cleaning compounds
Lime wastewater	Sludge acid
Lime and water	Spent acid
Spent caustic	Spent mixed acid
	Spent sulfuric acid
Potential consequences: Heat generation, violent reaction	
Group 2-A	**Group 2-B**
Asbestos waste and other toxic wastes	Cleaning solvents
Beryllium wastes	Data processing liquid
Unrinsed pesticide containers	Obsolete explosives
Waste pesticides	Petroleum waste
	Refinery waste
	Retrograde explosives
	Solvents
	Waste oil and other flammable and explosive wastes
Potential consequences: Release of toxic substances in case of fire or explosion	
Group 3-A	**Group 3-B**
Aluminum	Any waste in Group 1-A or 1-B
Beryllium	
Calcium	
Lithium	
Magnesium	
Potassium	
Sodium	
Zinc powder and other reactive metals and metal hydrides	
Potential consequences: Fire or explosion; generation of flammable hydrogen gas	
Group 4-A	**Group 4-B**
Alcohols	Any concentrated waste in Groups 1-A or 1-B
Water	Calcium
	Lithium
	Metal hydrides
	Potassium
	Sodium
	SO_2Cl_2, $SOCl_2$, PCl_3, CH_3SiCl_3
Potential consequences: Fire, explosion or heat generation; generation of flammable or toxic gases	

Table II, continued

Group A	Group B
Group 5-A	Group 5-B
Alcohols Aldehydes Halogenated hydrocarbons Nitrated hydrocarbons and other reactive organic compounds and solvents Unsaturated hydrocarbons	Concentrated Group 1-A or 1-B wastes Group 3-A wastes
Potential consequences: Fire, explosion or violent reaction	
Group 6-A	Group 6-B
Spent cyanide and sulfide solutions	Group 1-B wastes
Potential consequences: Generation of toxic hydrogen cyanide or hydrogen sulfide gas	
Group 7-A	Group 7-B
Chlorates and other strong oxidizers Chlorines Chlorites Chromic acid Hypochlorites Nitrates Nitric acid, fuming Perchlorates Permanganates Peroxides	Acetic acid and other organic acids Concentrated mineral acids Group 2-B wastes Group 3-A wastes Group 5-A wastes and other flammable and combustible wastes
Potential consequences: Fire, explosion or violent reaction	

[a]Source: "Hazardous Waste Management–Law Regulations and Guidelines for the Handling of Hazardous Waste," California State Department of Health (February 1975).
[b]Mixing a Group A waste with a Group B waste may have the potential consequence as noted.

Contingency Plan

Successful operation of a hazardous waste management facility relies on use of sound management techniques. An operational plant which provides for safe handling of the wastes must be developed, and adherence to that plan is mandatory. The operational plan should contain information indicating how the entire facility functions. This plan must be kept updated and current. Procedures indicating how hazardous wastes are processed, pretreated or disposed of at the facility and evaluation of the type of equipment employed should be included in this plan. The sequence of operations indicating flow schemes for processing, reclamation or disposal of various hazardous wastes and the proposed development stages of disposal areas or processing systems for different types of hazardous wastes should be outlined. Compatibility of various waste types should be evaluated to determine beneficial or detrimental combinations or mixtures of hazardous wastes [6].

Adequate supervision, trained personnel and effective monitoring at the facility must be provided. Facility operations must be supervised at all times during working hours by someone familiar with the dangerous nature of the various waste types. The name, address and telephone number of personnel in direct charge of the facility operations should be available to the regulatory agencies. A responsible person familiar with the facility and the preplanned emergency procedures should be available at all times. The supervisor must maintain the handling and treatment operations according to the operational plan. He must ensure that the facility is functioning in conformance with the regulations governing its operations.

Hazardous waste unloading areas should have wheel stops to prevent trucks from backing into waste material or containers. The wastes should not be allowed to accumulate where vehicles can come into contact with them. Unloading areas should be concrete pads with sloped areas to direct any spillage into proper disposal areas. Proper cleaning equipment should be provided at the facility for cleaning of spilled and splashed waste materials.

Hazardous wastes must not be stored in areas where such storage would be inconsistent with surrounding land use or interfere with the environment from unplanned release of such wastes. Hazardous waste must be stored in leakproof containers or tanks such that there is no discharge from the containers or tanks during their expected lifetime. The waste must not be stored for longer than the life of the container or tank. Containers and tanks of incompatible wastes must be separated. The storage sites must be designed, operated, located and constructed such that spills and leaks are contained within specific areas. All containers and tanks should be accessible and stored on concrete pads, wooden pallets or other means to prevent direct contact with the ground. Containers should be covered for protection from the weather and to prevent water from standing on or in them. All storage sites should be inspected daily by facility personnel. Leaks and spills should be reported immediately and containment and clean-up operations initiated as soon as possible. Records should be kept on all wastes, including information on waste types, characteristics, quantities, waste generators and the time and date the waste was received and put into storage.

Mitigating Damages

Hazardous waste materials may cause any number of environmental problems. Many times hazardous waste problems that do occur are not accidents at all, but instead they are the result of ignorance, carelessness or callousness. Improper management practices relative to hazardous waste transport, storage and disposal have resulted in significant, adverse impacts on surface and groundwater quality. EPA has published hazardous wastes disposal damage reports which contain case history descriptions documenting groundwater contamination, contamination of house wells, personal and animal injury and death resulting from the intake of contaminated groundwater.

Accidents resulting from inadequate hazardous waste management can cause adverse environmental and health impacts. These impacts can be attributed to acute (short-range) or chronic (long-range) effects of the associated hazardous wastes. Improper management of hazardous wastes is manifested in numerous ways. Waste discharges into surface waters can decimate aquatic plant and animal life. Contamination of land and groundwaters can result from improper storage and handling techniques. transportation accidents or indiscriminate disposal.

In assessing the impact of improper hazardous waste management practices, several factors merit emphasis. Because of the nature of hazardous materials, impacts will normally be more severe and persistent due to the potential toxic nature of most hazardous waste. Lack of or reduced rates of degradability and requirements of greater dilution factors to mitigate adverse effects become important. Due to the persistence of most of the hazardous waste materials, a corrective program for recovering the contaminated material becomes essential. Usually, a costly and lengthy contaminant containment and recovery program is required. Permitting a waste to remain in the ground and migrate within the groundwater system to some point of discharge would generally have serious adverse environmental impact.

Reported causes of the most frequently occurring accidents have been found to be broken pipelines, leaking and overflowing tanks, loss of oil, faulty pressure gages, faulty valves, pump failures, corrosion in lines, dismantling of equipment and plumbing connection errors where multiple pipelines are involved. At hazardous waste management facilities, bypasses and overloads can cause unnecessary accidents and pollution problems. The owners and operators of hazardous waste management facilities are responsible for inspection of storage and processing sites daily. They are responsible for reporting leaks and spills, for containment and clean-up of those leaks and spills and keeping records of such incidents.

When accidents or incidents occur, state governments have a major role to play in responding to those accidents. State governments have the responsibility to require adequate advance planning to prevent accidents from happening and to enable the responsible party to take immediate corrective action. Adequate training should be provided for personnel who handle hazardous materials. The states must do everything possible to mitigate damages when accidents occur. The states should develop the specialized knowledge necessary to cope with accidents involving hazardous materials or have ready access to people who possess this knowledge. A possible source of information to the state governments is the chemical industry, which may have available expertise in the handling and disposal of hazardous materials.

A major role of state governments in damage mitigation is in communications. An emergency telephone network in regional offices and in a centralized area should be maintained. Answering services by 24-hr/day operators, fully informed on whom to contact and how to contact them, should be provided since accidents may occur at night, on weekends or on holidays.

Of equal importance is the role of informing the public through the news media of the status of clean-up efforts and potential danger. Keeping the problem in proper perspective is one of the prime concerns. The states should have the capabilities for sampling, testing, and monitoring and assessing the extent of damages or the potential for damages. A prime function during clean-up should be the identification and location of sites or facilities that can handle the hazardous material that must be removed from the accident area. Long-range monitoring should also be performed to assure that the clean-up efforts adequately handle the problem. One of the tasks of the state government should be to assess fines and civil penalties after an investigation of the accident where the laws provide for such penalties.

Congress has declared that it is the policy of the United States that there should be no discharge of oil or hazardous substance into or upon the navigable waters of the U.S., adjoining shorelines or into or on the waters of the contiguous zone (National Oil and Hazardous Substances Pollution Contingency Plan). The primary thrust of this plan is to provide a coordinated federal response capability at the scene of an unplanned or sudden, and usually accidental, discharge of oil or hazardous substance that poses a threat to the public health or welfare.

Each of the primary and advisory federal agencies has responsibilities established by statute, Executive Order or Presidential Directive which may bear on the federal response to a pollution discharge. The plan intends to promote the expeditious and harmonious discharge of these responsibilities through the recognition of authority for action by those agencies having the most appropriate capability to act in each specific situation.

RESPONSE TO EMERGENCY SITUATIONS

Onsite alarm systems that can be activated quickly by plant operators to alert all plant personnel when an accidental release has occurred or is imminent are mandatory requirements for hazardous waste management facilities. The alarm system should be capable of alerting an emergency response official outside the perimeter of the facility when only a few employees are onsite, such as during hours when the facility is not operating or operating at a reduced level of effort.

A responsible official at the facility should be designated to make accurate assessment of the accident and emergency response. This assessment should include the following:

1. identification of the character and composition of the material that was released;
2. the volume, extent of discharge and expected direction and rate of travel;
3. the degree of hazard or imminent hazard and what action should be taken to mitigate damages;

4. which plant operations should be halted; and
5. what corrective measures may be necessary both at the facility and in the local community.

In conjunction with the assessment, an accident notification procedure should be implemented. A procedure list should be prepared which includes the names and telephone numbers of governmental, emergency, medical and public agencies that should be contacted during emergencies. These notifications should be documented with respect to person contacted, time and date of contact and information conveyed.

Procedures necessary to contain and recover spilled waste and to deal with fire and explosion should be included in the emergency response program. In the event of an accidental spill or release, the facility operators must respond immediately to restrict the movement of the waste from the spill site. Action should be initiated as soon as possible for collection of the spilled waste and any contaminated materials. The collected waste and contaminated material may require temporary deposition or storage in secure emergency storage areas. These materials should be treated as a hazardous waste and transported from the temporary storage area to approved treatment and/or disposal sites.

Fire and explosion are serious problems associated with many types of hazardous wastes. Factors that should be considered for inclusion in the fire prevention and control program include the following:

- portable fire extinguishers and control equipment;
- water available in adequate quantities and pressure;
- special extinguishing equipment using foam, inert gas or dry chemicals;
- regular maintenance, inspection and testing of fire control equipment;
- training and instruction of plant personnel on the proper use of this equipment;
- procedures for identification and isolation of wastes which might be flammable, explosive or reactive when subjected to fire;
- minimization of possible ignition sources by repair and inspection of machinery and equipment and by employing proper electrical equipment to prevent spark ignition or explosion; and
- smoking restrictions.

In every case of spillage or potential spillage of oil or hazardous substances to inland waters, notification must be given to EPA. EPA, through the Office of Water and Hazardous Materials, provides expertise on environmental effects of pollution discharges and environmental pollution control techniques, including assessment of damages. During emergency situations, the EPA also advises the Regional Response Team and On-Scene Coordinator of the degree of hazard a particular discharge poses to the public health and safety. When the On-Scene Coordinator receives a report of an emergency hazardous waste discharge, the following sequence of actions should be observed:

1. investigate the report to determine pertinent information, such as the threat to public health, type and quality of hazardous waste discharges, source of the discharges, etc;
2. notify appropriate agencies;

3. determine whether removal and corrective measures are being carried out properly; if not, determine need for additional measures;
4. designate the severity of the situation and determine the future course of action to be followed.

Various telephone hotlines exist for reporting emergency situations and providing immediate advice for dealing with emergencies associated with hazardous materials. The following agencies provide this type of assistance:

- U.S. Environmental Protection Agency
- U.S. Coast Guard
- Chemical Transportation Emergency Center (CHEMTREC), a public service of the Manufacturing Chemists Association
- U.S. Department of Transportation (DOT).

DOT has prepared a "Hazardous Materials Emergency Action Guide" to help emergency service personnel during the first 30 minutes of an incident involving a spill of volatile, toxic, gaseous and/or flammable material in bulk quantities [7]. General and specific safety procedures to follow are provided in spill guides arranged alphabetically for the hazardous materials listed in Table III. An example for nitric acid is presented in Figure 3.

DISCUSSION AND CONCLUSIONS

A monitoring program should be developed to detect problem areas associated with the operation of hazardous waste management facilities. Monitoring should be initiated before the start of operations at the facility to gather baseline data and information on environmental conditions before the startup of the facility. Effective monitoring can be accomplished by surveillance programs by regulatory agencies and by observation and sample collection by the facility operators. Groundwater, surface water, air quality and some biological sampling should be conducted to confirm the performance of the facility control measures. Site monitoring should be used to show that hazardous waste or dangerous derivatives are not escaping from the facility and that the protective or confinement barriers are effective. A systematic review should be established to determine if the control methods and programs are working effectively. Adequate records of the types, chemical composition, concentration and quantities of hazardous wastes received at the facility should also be maintained. Information on how the wastes were handled and treated and the locations where they were applied to the land should also be documented.

Hazardous waste handling, transport, treatment and ultimate disposal can result in many types of accidents. Emphasis should be placed on the objectives of waste handling at hazardous waste management facilities for the prevention and remedial procedures and methodologies for responding to accidents and spills at these facilities. These objectives include the following:

1. maintaining waste accountability;
2. employee safety and accident prevention;

Table III. Hazardous Materials

Acrolein	Hydrogen Sulfide
Acrylonitrile	Liquid Petroleum Gas
Ammonia	Methane, liquid
Ammonia, anhydrous	Methylamines, anhydrous
Boron Trifluoride	Methyl Bromine
Bromine	Methyl Chloride
Carbon Disulfide	Methyl Ethyl Ether
Chlorine	Methyl Mercaptan
Dimethyl Ether	Monomethylamine
Dimethyl Sulfate	Nitric Acid, fuming
Dimethylamine	Nitrogen Tetroxide
Epichlorohydrin	Oleum/Sulfur Trioxide
Ethyl chloride	Oxygen, liquid
Ethylene	Phosgene
Ethylene Oxide	Phosphorus Trichloride
Ethyleneimine	Propane/LPG
Fluorine	Sulfur Dioxide
Hydrocarbon Fuels	Sulfur Trioxide
(gasoline and similar fuels)	Titanium Tetrachloride
Hydrogen, liquid	Trimethylamine
Hydrogen Chloride	Vinyl Chloride
Hydrogen Fluoride	

3. handling and disposal guidelines for incompatible wastes;
4. development of effective operational and contingency plans; and
5. procedures for counteracting and mitigating damages.

Important design considerations for hazardous waste management facilities include security, access roads, unloading areas, waste classification facilities, processing and disposal systems, and drainage. All hazardous waste facilities, when closing, should terminate operations in accordance with their operations and contingency plans, to minimize accidents and incidents in the future. Maintenance and monitoring should be continued after abandonment of the facilities. Permanent records of hazardous waste types and their locations within the facility should be kept available. Problems noted by monitoring and surveillance programs should be corrected immediately.

Accidents resulting from hazardous waste management practices can cause adverse environmental and health impacts. Owners and operators of hazardous waste management facilities are responsible for inspection of

Nitric Acid, Fuming

(Oxidizer, Corrosive, Poisonous)

Potential Hazards

Fire: —May ignite combustibles.

Explosion: —Mixtures with fuels may explode.
—Runoff may create fire or explosion hazard in sewer system.

Health: —Contact may cause burns to skin and eyes.
—*Vapors may be fatal if inhaled.*
—Runoff may pollute water supply.

Immediate Action

—Get helper and notify local authorities.
—If possible, wear self-contained breathing apparatus and full protective clothing.
—Keep upwind and estimate *Immediate Danger Area.*
—Evacuate according to *Evacuation Table.*

Immediate Follow-up Action

Fire: —**Small Fire:** Dry chemical or CO_2.
—**Large Fire:** Water spray or fog.
—Move containers from area if without risk.
—Cool containers with water from *maximum distance* until well after fire is out.
—For massive fire in cargo area, use unmanned hose holder or monitor nozzles.

Spill or Leak: —Do not touch spilled material.
—Stop leak if without risk.
—Keep combustibles away from spilled material.
—Use water spray to reduce vapors.
—**Small Spills:** Take up with sand, earth or other noncombustible, absorbent material.
—**Large Spills:** Dilute with large amounts of water and dike for later disposal.
—Isolate area until gas has dispersed.

First Aid: —Remove victim to fresh air. Call for emergency medical care. *Effects of contact or inhalation may be delayed.*
—If victim is not breathing, give artificial respiration. If breathing is difficult, give oxygen.
—If victim contacted material, immediately flush skin or eyes with running water *for at least 15 minutes.*
—Remove contaminated clothes.
—Keep victim warm and quiet.

Figure 3. Emergency action

For Assistance Call Chemtrec toll free (800) 424-9300

In the District of Columbia, the Virgin Islands, Guam, Samoa, Puerto Rico and Alaska, call (202) 483-7616.

Additional Follow-up Action

—For more detailed assistance in controlling the hazard, call Chemtrec (Chemical Transportation Emergency Center) toll free (800) 424-9300. You will be asked for the following information:

- Your location and phone number.
- Location of the accident.
- Name of product and shipper, if known.
- The color and number on any labels on the carrier or cargo.
- Weather conditions.
- Type of environment (populated, rural, business, etc.)
- Availability of water supply.

—Adjust evacuation area according to wind changes and observed effect on population.

Water Pollution Control

—Nitric Acid is water soluble and can kill fish. Prevent runoff from fire control or dilution water from entering streams or drinking water supply. Dike for later disposal. Runoff to storm sewers or sanitary system is acceptable if a water deluge and/or flooding is possible. Notify Coast Guard or Environmental Protection Agency of the situation through Chemtrec or your local authorities.

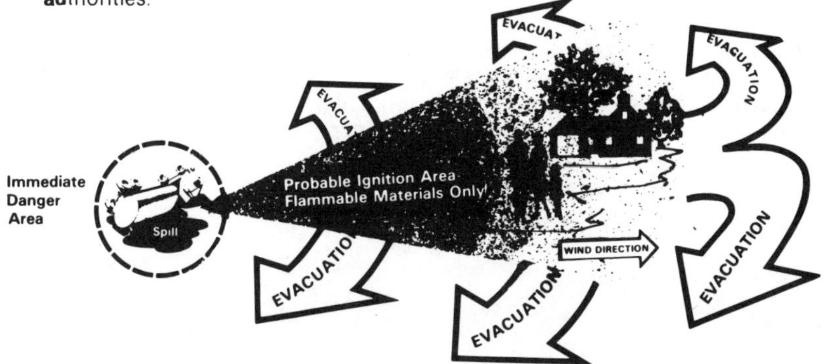

Evacuation Table — Based on Prevailing Wind of 6-12 mph.

Approximate Size of Spill	Distance to Evacuate From Immediate Danger Area	For Maximum Safety, Downwind Evacuation Area Should Be
200 square feet	65 yards (78 paces)	1,528 feet long, 1,056 feet wide
400 square feet	100 yards (120 paces)	4,224 feet long, 1,584 feet wide
600 square feet	120 yards (144 paces)	3,168 feet long, 2,112 feet wide
800 square feet	140 yards (168 paces)	3,696 feet long, 2,112 feet wide
In the event of an explosion, the minimum safe distance from flying fragments is 2,000 feet in all directions.		

guide for nitric acid.

storage and processing areas and for containment and clean-up of leaks and spills. State and federal governments as well as facility personnel have a major role to play in responding to hazardous waste accidents and incidents when they occur.

REFERENCES

1. Metry, A. A., and E. L. Stover. "Ultimate Disposal of Hazardous Solid Waste," paper presented at the Hazardous Solid Waste Management Seminar sponsored by the Pennsylvania Department of Environmental Resources, August 1976.
2. Manko, J. M., and B. S. Katcher. "Hazardous Wastes–From Cradle to Grave and Beyond," *Compost Sci./Land Utilization* (July/August 1978), pp. 24-26.
3. Jepson, H. G. "Coping with Results of Mercury and Arsenic Spills," *Poll. Eng.* (Oct. 1978), pp. 57-60.
4. Metry, A. A., Ed. *The Handbook of Hazardous Waste Management* (Westport, CT: Technomic Publishing Co., Inc., in press).
5. "Hazardous Waste Management–Law, Regulations and Guidelines for the Handling of Hazardous Waste," California State Department of Health (Feburary 1975).
6. "Guidelines for Hazardous Waste Land Disposal Facilities," California State Department of Health (January 1973).
7. "Hazardous Materials–Emergency Action Guide," U.S. Department of Transportation, National Highway Traffic Safety Administration (1976).

CHAPTER 3

UTILITY INDUSTRY RESEARCH ACTIVITIES FOR THE IDENTIFICATION AND DISPOSAL OF HAZARDOUS WASTES

Dean M. Golden
Electric Power Research Institute
Palo Alto, California

INTRODUCTION

The Electric Power Research Institute (EPRI) was founded in 1972 by the nation's electric utilities to develop and manage a technology program for improving electric power production, distribution and use. Over 500 utility companies, representing over 80% of installed generating capacity are contributing members. Currently, there are over 900 ongoing projects with a total funding level in excess of $1 billion.

The purpose of this chapter is to describe the recently completed and ongoing EPRI research activities relating to the identification, characterization and disposal of hazardous wastes.

As a natural consequence of power plant operation, a diverse set of waste streams or by-products are created. These solid/liquid by-products include coal fly ash, bottom ash, flue gas desulfurization (FGD), solid wastes (scrubber sludges), mill pyrite rejection, coal/limestone pile runoff, makeup water treatment blowdown, equipment drainage wastes, cooling tower sludges and blowdown, and other miscellaneous streams. The disposal of these liquid and solid wastes is generally approached as a problem requiring an integrated solution through cascading reuse of water at stepwise lower levels of water quality requirements, and the transfer of solid waste toward a single ultimate treatment and disposal process. The electric utility industry currently uses pits, ponds and landfill disposal sites costing about $2/ton (dry basis). By 1985, it is estimated that there will be 200 million ton/yr of these wastes from the utility industry.

Those modest prices will soon climb by at least an order of magnitude. Under the Resource Conservation and Recovery Act of 1976 (RCRA), the U.S. Environmental Protection Agency (EPA) has proposed strict new regulations for the identification, treatment, storage and disposal of hazardous solid wastes. This includes a series of tests for determining whether a waste is to be classified as hazardous or not. An EPRI data study [1] published in September 1978 indicates that ash, scrubber sludge, metal cleaning wastes, boiler blowdown and other power plant wastes might approach or exceed certain of the draft EPA criteria, specifically in the areas of toxicity and corrosiveness, and therefore be considered hazardous.

In addition to the by-products of power production which may be considered as hazardous, another waste already identified as hazardous with a resultant impact on utilities is polychlorinated biphenyls (PCB). More than half of the estimated 750 million lb of PCB are in utility system equipment, almost all in transformers and high voltage capacitors. PCB are regulated as hazardous wastes under the Toxic Substances Control Act of 1976 (TSCA) and RCRA. EPA regulations prescribe high-temperature incineration as the principal method of disposition of PCB with certain items allowed to be placed in chemical waste landfills.

EPRI SOLID WASTE DISPOSAL RESEARCH OBJECTIVES

In response to utility industry needs in the disposal of hazardous and nonhazardous solid wastes, EPRI has prepared a major interdepartmental solid waste disposal program. The new program will assist the industry in reducing the cost of disposal. The multimillion-dollar program has five main objectives:

1. Define the physical and chemical nature of solid wastes. EPRI efforts will include evaluating EPA test procedures and determining the reproducibility of test results. EPRI will also look into fuel composition, sample variability, surface physics and chemistry, leachate chemistry, and transport phenomena.
2. Develop an economic and environmentally sensitive method for assessing the hazard potential of utility wastes. This will include health effects and biomedical studies, with studies on trace element toxicity and an assessment of the Ames mutagenic assay.
3. Develop resource recovery processes and utilization systems. EPRI will develop ways to remove hazardous constituents from waste treatment. FGD sludges and advanced system residues will also be studied. A handbook for using fly ash as structural fill will be developed, and the use of fly ash to enhance methane production will be evaluated.
4. Develop safe solid waste disposal systems. EPRI will study FGD sludge and ash disposal and will develop sludge and ash disposal handbooks. Disposal site studies are proposed, including groundwater modeling of leachates, an assessment of the environmental effects of current disposal systems and a study of the effectiveness of various liners. Ocean and deep-mine disposal options will be explored, together with disposal methods for advanced system wastes, such as coal liquefaction and gasification wastes.
5. Respond to socioeconomic concerns facing the industry. EPRI will assess the economic impact of RCRA on the utility industry, as well as the applicability of RCRA hazardous waste regulations to the industry.

IDENTIFICATION OF HAZARDOUS WASTES

The "$64 question" these days in the utility industry is what power plant wastes may be classified as hazardous under screening test procedures established in RCRA regulations. The proposed Section 3001 regulations issued in December 1978 [2] establish a set of criteria to be used by a waste generator to determine whether a particular waste will be considered hazardous in the regulatory sense of the word. These criteria have been divided into six categories: ignitable, reactive, infectious, corrosive, radioactive and toxic. In the December 1978 proposed regulations, no utility solid wastes were specifically listed on the Section 250.14 (a) or (b) hazardous waste list, although many other industrial process wastes almost identical to utility wastes were included. The major area of concern is the toxic category which has the greatest potential applicability to fossil fuel power plant waste streams. Most of the toxicity criteria require long-term testing, using specific testing methods which are described in detail in the December 1978 regulations [2]. Several of these methods have been developed only recently, or are still in the development stage, so few data are available. In any case, there seems to be considerable question in the scientific community concerning the reliability of the proposed tests as an indicator of "hazardousness."

Under the proposed EPA regulations, the individual waste generator will be responsible for determining if the waste it produces is hazardous. Although a waste must be declared hazardous if it gives a positive result in just one of the tests for toxicity, corrosiveness, radioactivity, reactiveness, ignitability or infectiousness, EPA has created a new class of wastes called "special wastes" which include scrubber sludges, fly ash and bottom ash which are otherwise found to be hazardous. For these three utility wastes, the EPA has postponed the stringent rules relating to the storage and disposal of hazardous wastes. During the interim period until performance standards for "special wastes" are adopted, minimum security measures and full monitoring requirements have been proposed. The validity of this new category of "special wastes," which was established because of the economic impracticality of requiring power plant by-products to be disposed of as hazardous wastes, will undoubtedly be questioned by environmental organizations on one hand and by other industries seeking similar classification. At this point, it is unclear what happens when special wastes are combined with low volumes of hazardous wastes for disposal or if lime (a corrosive substance by virtue of its pH) is added to stabilize a special waste. If one were to use the logic of the Water Pollution Control Act regulations, when a hazardous waste is combined with a special waste, the entire waste stream would have to be treated as if it were hazardous.

EPRI SOLID/HAZARDOUS WASTE DISPOSAL RESEARCH

For the past year, EPRI has carried on a program in the identification, characterization and disposal of utility solid/hazardous wastes. This was undertaken in response to industry concerns that certain power plant wastes might be identified as hazardous under the evolving EPA criteria. An EPRI data study published in September 1978 [1] did indicate that ash, scrubber, sludge, metal cleaning wastes, boiler blowdown and several other power plant wastes might approach or exceed certain of the draft EPA criteria, particularly in the areas of toxicity, corrosiveness or radioactivity. It should be noted, however, that these studies were based on worst-case conditions with sample data taken from raw waste streams rather than elutriates obtained from the EPA extraction procedure.

Identification and Characterization Research

One portion of ongoing work within the EPRI Energy Analysis and Environment Division, in the Physical Factors Program, is a project to assess the reproducibility of the EPA tests proposed for use in deciding if ash is a hazard or not. It has been shown, based on these leaching tests results, that the complex mixture of trace elements that make up coal ash vary greatly depending on circumstances. Fly ash constituents, for example, differ according to type of fuel used, method of combustion and sampling location. Any test which will be used to decide whether a waste is hazardous will have considerable economic importance to the waste generator. There can be as much as an order of magnitude difference in disposal costs, so the regulatory screening test should at least meet the standards of precision required of ASTM standards.

A project to study the statistical variation of the trace components in ash began early in 1979 in conjunction with a study evaluating the chemistry of trace metals in ash.

Health Effects of Utility Waste Products

The Health Effects and Biomedical Studies Program within the Energy Analysis and Environment Division has placed its main emphasis on investigating the possible health effects of combustion of fossil fuels. Both laboratory and epidemiologic techniques of study have been used in attacking this problem.

As a result of rising concern in government and among members of the public, as well as the utility industry, an expanded program is being planned which will allow the assessment of selective hazards within the utility power plant workplace. This will include studies of radiation, noise, chemical toxicity and carcinogenesis.

Results expected during 1979 from this research program include: (1) publication of methods of application of fly ash testing of *in vitro* systems; (2) publication of an assessment of the predictive potential of the Ames bacterial mutagenic assay; and (3) a report assessing the carcinogenic risk attributable to fossil fuel combustion and conversion processes.

The proposed Section 3001 regulations are contained in "Advance Notice of Proposed Rulemaking" for the establishment of regulations on required toxicity tests for mutagenic activity. The application of these sophisticated laboratory procedures to the screening of wastes to determine their "hazardousness" will undoubtedly increase research activity in this area.

Resource Recovery and Utilization Research

The percent use of fly ash, bottom ash and boiler slag is currently about 10%, 25% and 40%, respectively [3]. One side effect of a waste being declared hazardous is that it will seriously hamper (or eliminate) the use of these useful combustion by-products and therefore increase the volume of wastes that require disposal. For this reason, the emphasis of the EPRI Interdepartmental Solid Waste Program has been on disposal rather than utilization, at least until the regulatory "dust settles."

One project (RP1170), in the Energy Analysis and Environment Division, prepared a "Fly Ash Structural Fill Handbook" as a means of encouraging this use. This handbook was published in spring 1979.

A two-year project (RP1404-2) in the Fossil Fuel and Advanced Systems Division, started in December 1978, is studying methods for extraction of the trace metals in fly ash. Since these are the same trace metals which can make an ash "hazardous," their removal before disposal will have important environmental and economic advantages. Many of the metals recovered have a significant commodity value. The project will result in the identification of the most promising removal process including process flowsheets and detailed designs for a demonstration plant, as well as cost estimates and expected benefits from the recovered resources.

At this time, the potential for any large-scale by-product recovery from nonregenerable scrubber systems appears small [4]. One area of potential use of sludge is for reclamation of mines as a means of abatement of acid mine drainage. It is hoped that a mine disposal demonstration will be approved for an EPRI project later in 1979. An assessment of the impact of the regulations issued by the Office of Surface Mining pursuant to the Surface Mining and Reclamation Act of 1977 on disposal of utility wastes is underway. If the assessment indicates that mine reclamation using scrubber sludge and ash by-products is feasible from a regulatory standpoint, a mine "disposal" demonstration project can be developed.

Hazardous Waste Disposal Research

Until the EPA regulations under Section 3001 are made final, the only bona fide hazardous wastes with which the industry msut contend are PCB, which are regulated under TSCA as well as RCRA.

The EPRI PCB project was authorized by the Board of Directors in February 1978. It is jointly funded by the Energy Analysis and Environment Division and the Fossil Fuel and Advanced Systems Division. The project reports were released sequentially as a five-volume set during 1979.

PCB Disposal Research

The objective of the EPRI research project (RP1263) on PCB disposal is to develop guidelines for the evaluation of acceptable alternative disposal methods for wastes containing PCB. The intent of this project is to assist electric utilities, which possess more than half of the PCB manufactured between 1929 and 1977 (when production ceased). The term PCB designates a group of synthetic chlorinated organic compounds first introduced in large-scale use in 1929. Although there are 209 different chlorinated biphenyls which are collectively called PCB, commercial PCB are generally mixtures of 100 different isomers.

The EPRI guidelines to be provided in Volume 1 on PCB disposal are presented in the form of a data base, including relevant information on the following:

1. PCB production and use;
2. PCB disposal regulations (including proposed regulations);
3. projected regional PCB disposal requirements;
4. available PCB incineration technology and proposed commercial facilities; and
5. PCB landfill design and available commercial facilities.

The general conclusions and findings of this EPRI research study are as follows:

1. By using industry equipment service life data and PCB equipment distribution by region, it is apparent that significant PCB disposal capacity will be required in all parts of the United States for the next 40 years.

2. These is insufficient landfill capacity at present for PCB solid waste, but by next year, it may be sufficient.

3. The incineration capacity for PCB solids is presently zero and will not be available in sufficient capacity by the required January 1, 1980 deadline. The lack of available commercial PCB incineration is due to stringent design criteria which cannot be met by any existing facilities, the lack of demonstrated demand for this capacity (until the effective date of the incineration regulations) and, most importantly, the public opposition to the siting of new facilities.

4. The technology for PCB incineration is available with utility boilers, cement kilns and several incinerator configurations apparently meeting the EPA specified criteria. Environmentally acceptable disposal of PCB wastes necessitates the consideration of several factors not usually encountered in chemical waste incineration. As a result of the high stability of the PCB molecule, relatively high temperatures and residence times are required for complete destruction. In addition, the chlorine present in PCB can form extremely corrosive substances in the exhaust gas. Based upon full-scale test burns on utility boilers, solids/liquids incinerators and cement kilns, it has been demonstrated that adequate destruction of PCB materials can be achieved.

5. The EPRI PCB project includes the development of conceptual designs for the systems required for the incineration of PCB solids and liquid

mixtures, along with cost estimates. In general, the unit costs for PCB liquids and solids disposal in a cement kiln are less than half that of a utility boiler disposal system due to the differing feed rates allowable.

6. The recommended actions for utilities to ensure that there will be adequate disposal capacity include: (1) utility development of new integrated incinerator systems specifically for PCB combustion; (2) adaptation of existing boilers for PCB, (3) utility-owned cement kilns for PCB disposal and use of other solid waste by-products such as fly ash and gypsum; (4) establishment of utility-owned hazardous waste landfills; and/or (5) use of existing commercial facilities in those regions of the country with sufficient capacity.

7. The most important consideration in evaluating the alternative PCB disposal options is the cost of transportation to the disposal facility. The cost of PCB transport is approximately $0.15/ton-mi.

8. Volumes II and III of the EPRI PCB project reports relate to the development of exposure and contamination control (ECC) plans to ensure that the risks associated with PCB activities are minimal. Model operation plans addressing assembly and servicing techniques, spill-free use of equipment and containment procedures for prevention of accidental releases are included in these EPRI reports.

9. Volumes IV and V of the EPRI PCB project reports (issued later in 1979) will document the test incineration of PCB solids and liquids in an integrated commercial incinerator and the incineration of PCB contaminated mineral oil in an oil-fired utility boiler.

Solid Waste By-Product Disposal Research

Regulatory Considerations

Unfolding EPA regulations on what constitutes a nonhazardous, hazardous or special waste create considerable uncertainty in designing a research program in by-product disposal because of the differing performance standards. Before reviewing the EPRI program, a brief summary of the proposed design and operational constraints for these three categories of wastes is necessary.

The proposed design and operational criteria for nonhazardous wastes (issued in February 1978) [5] were expected to be finalized in Spring 1979. These federal guidelines are for the purpose of protecting surface and groundwaters. Although they are guidelines, many states view them as standards. The guidelines for siting, leachate management and control include:

1. The fill bottom must be above the seasonal high groundwater table.
2. No direct connection between fill and surface water is allowed.
3. Runoff must be diverted from the fill site with structures capable of containing a 25-year flood.
4. A cap of low permeability (6 in. of clay), overlain with 18 in. of topsoil, should be used to allow the establishment of natural vegetation.
5. If the groundwater under a disposal site is designated for human consumption or,

if undesignated, contains less than 10,000 mg/l total dissolved solids, an impermeable liner must be provided. Natural liners must be at least 1 ft thick with a permeability of 1×10^{-7} cm/sec or less. Artificial liners must be at least 20 mils thick and durable enough to last for the design life of the facility.

6. The siting criteria for new nonhazardous waste disposal areas specify that the disposal facilities not be located in environmentally sensitive areas (wetlands, 100-year floodplains, critical habitats, areas of active faults, etc.).

The proposed design and operational criteria for hazardous wastes (issued in December 1978) [2] were expected to be finalized in December 1979. A federal permit is required for the disposal operation if the waste is classified as hazardous. In general, the hazardous waste regulations are more stringent than regulations governing nonhazardous or special waste disposal in that they *require* disposal in secure sites which do not allow interaction with the environment. The standards for siting, leachate management and control include:

1. The siting constraints for hazardous wastes include the restrictions for nonhazardous waste (with exception of permafrost and agricultural lands) and several new ones. The added restrictions prohibit locating in coastal high-hazard areas and within a 500-year floodplain. (This could open a whole new market for ark builders for disposal facilities!) Additionally, a 200-ft buffer zone is required between active positions of the disposal area and the property line.

2. Landfills over usable groundwater are permitted no direct contact between the landfill and surface or groundwater. Sufficient depth must be provided between the landfill bottom and groundwater to allow for leachate monitoring and adequate time for remedial measures to correct any future problem.

3. The wastes must be treated to reduce water content, solubility and toxicity.

4. Where natural conditions allow, a liner must be provided. Natural soil (clay) liners must be at least 10 feet thick with a permeability of 1×10^{-7} cm/sec or less.

5. Where natural conditions do not allow a natural liner to be formed, then a liner must be installed to: (a) allow containment and removal of leachate from below the fill; (b) provide a liner which is equivalent to 5 ft of soil (clay) with 1×10^{-7} cm/sec permeability; and (c) provide a drainage blanket on top of the liner to allow gravity drainage to a collection sump.

6. For ponds, liners must prevent seepage, both laterally and vertically. A double liner is required when natural conditions do not provide the required impermeability. Artificial liners must be on a stable base and be able to last at least 75% longer than the design life of the facility, as well as withstand degradation due to freeze-thaw.

For the three utility wastes (scrubber sludges, fly ash and bottom ash) categorized as "special wastes," if they are found to be hazardous by the screening tests, the stringent rules above have been suspended. In their place, the EPA has proposed regulations which provide for general protection of public health and the environment and require gathering additional information on these wastes. In effect, EPA has postponed the regulations on these three utility wastes until more data are gathered and evaluated. During this interim period, these special wastes are required to comply with (1) general facility standards; (2) general site selection criteria for new facilities; (3) security measures to limit access to the site, (4) record-keeping and reporting measures; (5) visual inspections; (6) closure and postclosure requirements; and (7) groundwater and leachate monitoring to develop data related to the environmental effects of disposal.

EPRI Disposal Research

An early product of the EPRI disposal research program was the flue gas desulfurization (FGD) sludge disposal manual [4], developed under RP786-1 and issued in February 1979. The manual considers the disposal of the wastes from lime, limestone, alkaline fly ash and double-alkali throwaway scrubbing systems designed primarily for flue gas desulfurization on utility boilers. The objective of the manual is to provide information and direction to utilities which can be used to select and design FGD by-product disposal systems and components. The manual contains chapters on current disposal practices, how to estimate waste composition and quantities, disposal alternatives, site selection, leachate, disposal area design, sludge processing, forced oxidation, thickening, dewatering, fixation/stabilization, transportation, costs and utilization.

The manual reviews the components of sludge along with their effects on processing and disposal. Three distinct methods are presented for estimating the quantities of sludges which will be produced under sets of assumed conditions. The options available for processing and disposal and the question of fixation versus stabilization are considered in light of the current regulations and the possible requirements to be established in the future under RCRA. The terms fixation and stabilization are often used interchangeably in the technical literature so there is no generally accepted difference between these terms. In the EPRI sludge disposal manual, a distinction is made for the sake of clarity. Stabilization is used to mean any chemical and/or physical treatment designed to improve chemical and/or physical properties. For example the addition of fly ash, soil or other material to the sludge to induce physical changes is considered stabilization. Fixation on the other hand is a form of stabilization which involves the addition of reagents which cause a chemical reaction in the sludge. The prospects for utilization were considered, and it is concluded that it is not likely that there will be any large-scale use. The various factors affecting cost are reviewed and procedures are recommended for estimating the components of total cost under differing assumptions.

In order to keep this sludge disposal manual useful in the years to come, it will be updated frequently to incorporate the latest technology or respond to the final EPA regulations issued to implement RCRA. The first revision was in late Spring 1979.

Although the sludge disposal manual includes sections on the disposal of fly ash when it is collected in the scrubber and becomes a part of the sludge or when it directly related to sludge disposal, the manual does not address the collection and disposal of fly ash as a separate by-product.

A separate ash disposal manual was prepared under RP1404-1. The purpose of this manual is to provide information in detail about fly ash disposal for use by utility design staff for guidance in the evaluation of technical and economic factors governing the selection of optimal disposal systems and locations. This manual will include: (1) detailed site-selection information including physical, engineering, regulatory, environmental and economic considerations; (2) information on the physical, chemical and engineering properties of ash and its leachate; (3) a summary of current disposal practices and an assessment where they may be deficient when compared to proposed regulatory criteria; (4) details of the design features, equipment selection, licensing and specific procedures necessary for the construction of new facilities which meet the regulatory criteria; (5) information to allow prediction of such factors as waste quantity, waste characteristics and system costs; (6) explanations of monitoring and monitoring well systems, including their costs; (7) information on site reclamation procedures for ash disposal areas; and (9) cost estimates and cost curves or tables for making preliminary, general level cost estimates.

The largest project in the EPRI by-product disposal program from a funding standpoint is a sludge disposal demonstration project at a 20-MWe limestone dual-alkali scrubber (RP1405). This project at the Scholz Power Plant of Gulf Power Company is being operated in conjunction with the EPA process evaluation demonstration. The demonstration and monitoring of sludge disposal at this experimental facility will provide a documented technical basis for the future design of full-scale disposal facilities. This project is an opportunity to identify and solve, on an experimental facility of tractable size, the potential engineering, operational and environmental problems associated with the disposal of high-sodium, high-sulfite sludges. The 20-MW demonstration size is sufficiently large so that scale-up to a full-sized facility can be performed. During the project, particular attention will be given to the following issues: (1) the difficulties, if any, associated with the mixing of sludge and ash on dewatering, handling, stability, leachability, etc.; (2) the techniques, costs, etc., for fixation and stabilization; (3) the site preparation requirements; (4) the composition of leachate, supernatant and runoff; (5) the rate of contaminant leaching and its composition; (6) identification of potential problems with recycling runoff and drainage from the landfill to the scrubber for makeup water; (7) problems, if any, associated with the high-sodium, high-sulfate characteristic of these sludges; and (8) long-term maintenance requirements and site reclamation criteria.

The project began in January 1979 and is 31 months in duration, although most of the demonstration was completed during 1979. Each of the three landfill disposal areas will be filled with a different material involving different mixes of fly ash, FGD waste and lime. The study in later years will evaluate the chemical and physical fate of the waste materials in the disposal site.

EPRI project RP1406 is a three-year project at the Conesville Station of the Columbus and Southern Ohio Electric Company. This is a monitoring and model development project. The site was selected because it is the first full-scale system utilizing the disposal of sludge treated by the IU Conversion System (IUCS) proprietary process. Since this is a first-of-a-kind operation, industry-sponsored surveillance was considered advisable to determine if full-scale application of this proprietary fixation process (1) reflects laboratory and test-pond results claimed by the proprietor; (2) provides an environmentally acceptable disposal method; (3) creates operating problems for the utility; and (4) meets criteria established by regulatory agencies.

A separate contractor is developing a model for predicting the quality and quantity of leachate and its migration path in the disposal area. The monitoring program will verify the predictive model since the monitoring well locations were based on the model.

One question often raised by utility design engineers in developing waste disposal facilities is what information is available on the long-term stability of sludge/ash mixtures. In response to this need, EPRI funded a project (RP1260-1) to investigate the long-term stability of sludge/ash mixtures. This laboratory investigation includes: (1) evaluation of long-term strength behavior, (2) long-term permeability studies, (3) long-term leachate potential, and (4) further study of the mechanisms (both chemical and physical) responsible for the changes in strength behavior. Approximately 250 test cylinders of different fly ash and sludge mixtures have been tested at 58 days, 500 days and will be tested again at two years. The results, published in Fall 1979, should be valuable in describing the short- and long-term mechanisms responsible for the development of physical strength.

The vast majority of existing utility industry solid waste disposal sites are lined with native soils rather than synthetic membranes. It is expected that this will change in view of the proposed stringent federal regulatory requirements (Section 3004 of RCRA), and increased environmental awareness on the part of industry and general public will create an impetus for new approaches to utility waste disposal. Soil containment may still remain the most prevalent design for utility waste disposal, but admixed materials and synthetic membranes will find increasing use as a leachate control technique. In environmentally sensitive areas such as floodplains and wetlands (where most power plants are located), the use of lining material may be the rule rather than the exception.

In view of this need, a new EPRI project (RP1457) was initiated in Spring 1979 to evaluate the effects upon a selected group of 14 liner materials to exposure to 9 types of potentially hazardous utility wastes over an extended period of time. The objectives of this laboratory study are to: (1) determine the durability and cost-effectiveness of using synthetic membranes, admixed materials and natural soils as liners for waste storage and disposal areas; (2) estimate the effective lives of liner materials exposed to different types of utility wastes under conditions which simulate those encountered in holding ponds, lagoons and landfills; and (3) develop a method for assessing the relative merits of the various liner materials for specific applications and for determining their service lives.

A separate part of this research project will prepare a state-of-the-art investigation into the groundwater monitoring systems to review current practices and provide guidelines for the proper design, location, construction and maintenance of these sytems. This phase of the project was published in late 1979.

Regulatory Impact Assessment Research

Although it is not the purpose or objective of the EPRI research program to review each piece of legislation or regulation which may affect the utility industry, occasionally a law will have a major impact on the way utilities will do business in the years ahead, and research studies will be performed. The Clean Air Act and RCRA are very significant in their impact on power plant design and operation. The solid waste disposal program at EPRI has performed one study [1], published in September 1978. During 1979, additional studies were planned to assess some of the implications of the regulations being issued pursuant to RCRA, as well as those issued under the Surface Mining and Reclamation Act of 1977.

CONCLUSIONS

This paper has presented an in-depth description of the recently completed and ongoing EPRI research activities relating to the identification, characterization and disposal of utility "hazardous" wastes. This overview of the EPRI interdepartmental solid waste disposal program has been interwoven in the present regulatory framework under RCRA as it affects utility by-product disposal operations. This EPRI research and development effort has combined the resources of the Water Quality Control and Heat Rejection Program of the Fossil Fuel and Advanced Systems Division, and the Biomedical Studies, Ecological Effects, Health Effects and Physical Factors Programs of the Energy Analysis and Environment Division. It is anticipated that the results of this multifaceted research program will make a contribution to the data base needed by the utility industry and policy-making bodies in energy and environmental decision-making. Much of the information is also transferable to other industries seeking to survive in the ever-changing regulatory environment.

REFERENCES

1. Hart, Fred C., Associates, Inc. "The Impact of RCRA (PL94-580) on Utility Solid Wastes," Electric Power Research Institute RP-878 (August 1978).
2. Environmental Protection Agency. "Proposed Rules under Sections 3001, 3002, and 3004 of the Resource Conservation and Recovery Act," *Federal Register*, 43(243) (1978).
3. De Carlo, V. A. et al. "Evaluation of Potential Processes for the Recovery of Resource Materials from Coal Residues: Fly Ash," Oak Ridge National Laboratory, TM-6126 (March 1978).
4. Baker, Michael, Jr., Inc. "FGD Sludge Disposal Manual," Electric Power Research Institute, FP-977 (January 1979).
5. Environmental Protection Agency. "Proposed Rules for Classification of Solid Waste Disposal Facilties," *Federal Register*, 43(25) (1978).

CHAPTER 4

THE ROLE OF THE SECURE LANDFILL IN HAZARDOUS WASTE MANAGEMENT

Edward R. Shuster and Louis E. Wagner

CECOS Inc.
Niagara Falls, New York

In August 1976 Congress enacted the Resource Conservation and Recovery Act (RCRA), which for the first time established federal control over solid waste disposal. The definition of "solid waste" in that law includes "solid, liquid, semisolid, or contained gaseous material" plus nearly any other waste not already controlled by the Clean Water Act or the Atomic Energy Act, essentially "by difference."

The subcategory of "hazardous waste" defined in RCRA is:

> . . . a solid waste, or combination of solid wastes, which because of its quantity, concentration, or physical, chemical, or infectious characteristics may–
>
> (A) cause, or significantly contribute to an increase in mortality or an increase in serious irreversible, or incapacitating reversible, illness; or
>
> (B) pose a substantial present or potential hazard to human health or the environment when improperly treated, stored, transported, or disposed of, or otherwise managed.

The key words for the purpose of this chapter are "when improperly managed." These are the materials that are addressed in this chapter, and a method for proper management.

The Proposed Regulations issued December 18, 1978, and the Final Regulations governing PCB waste disposal issued February 17, 1978 under the Toxic Substances Control Act (TSCA) are presently used as regulatory models. Within these regulations are described three different designs which, if followed, would lead the builder to an approvable chemical waste landfill. Included in the criteria are site-selection parameters, soils, hydrology, permeability, natural and manmade liner criteria, a sorting of types and classes of appropriate and inappropriate waste materials for this mode of disposal, synergism effects of one waste on another, cover material, method of covering used, leachate collection and treatment, closure or sealing when the facility is full, postclosure care and maintenance, and a monitoring requirement during and after use of the site. These are all appropriate considerations, and, as might be imagined, will result in many more than three basic designs, depending on where the facility is located. Obviously a desert scenario has far different implications than an area of high precipitation.

The consideration of these criteria, an insistence on positive control and safety, and the application of chemical and physical treatments to reduce or eliminate the hazardous or toxic properties of wastes and immobilize them before placement in the chemical waste landfill has resulted in what is termed "secure landfill."

Secure landfill, then, is necessary and has been established as an equally viable and highly professional weapon in the hazardous waste management arsenal, alongside recycling, resource recovery, source radiation and incineration. If one considers material balance in each of those processes, the common denominator is that something is always left over.

The key for determining and assuring that a chemical waste landfill is a secure landfill, given a proper receptacle, is in choosing and conditioning the wastes properly such that their properties will either not change or will improve with time. This is an extremely important consideration because the technology, installed capacity and/or economics simply do not exist at the present for alternative methodology as applied to many wastes.

To determine if a waste is acceptable, it is necessary to obtain a representative sample plus a fairly complete compilation of total composition (both hazardous and nonhazardous species) and the physical and chemical properties of the waste. A laboratory-scale evaluation of treatment alternatives to condition the waste is performed. Techniques currently employed to pretreat wastes include neutralization, solidification in the form of a gypsum or mortar-type material, dewatering, adsorption, polymerization and buffering. It is pointed out that many wastes are treatable at the point of generation to permit direct placement in the secure landfill on arrival at the disposal site.

The NEWCO secure landfill system consists of a series of adjacent cells (Figure 1). Each cell is constructed to provide permanent containment using a triple-liner system which includes a reinforced polymer membrane between two layers of highly impermeable clay. Internal monitoring and collection wells are installed to collect incident rainfall during filling plus any dewatering of damp sludges. A convenient cell size is several acres, and depth is typically 30 ft plus a capping mound with a 1 on 3 grade to an additional height of 10-20 ft. Completed cells are capped off with additional clay or clay and membrane combinations, then topsoil and grass. A gas venting system under the cap ties into the collection pipes. External monitoring wells are provided to allow documentation of continued cell integrity

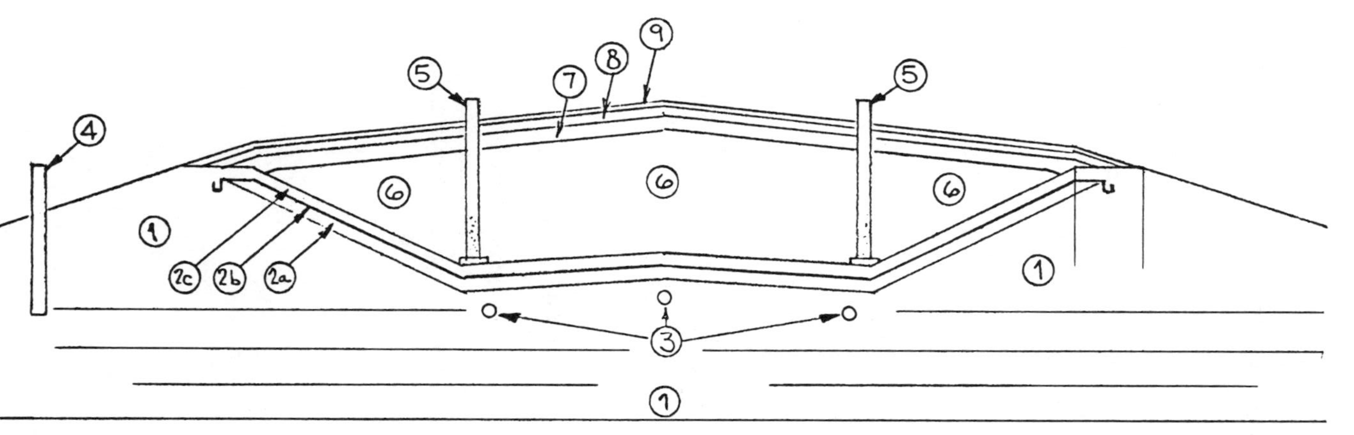

LEGEND

1. Impermeable clay-type sub soil
2. Triple liner system
   (a) compacted clay
   (b) reinforced polymer membrane
   (c) compacted clay
3. Under-drain monitoring system
4. External monitoring wells (3 ea.)
5. Internal monitoring wells (2 per cell), also leachate collection and removal for treatment
6. Internal volume for placement of approved wastes (pretreated for chemical and dimensional stability, carefully placed in a grid system according to engineering plan) and daily cover (includes manufactured components to buffer and stabilize matrix)
7. Soil cover
8. Impermeable clay cover
9. Topsoil and ground cover (grass)

Figure 1. NEWCO Chemical Waste Systems Inc. secure landfill cell.

through an ongoing sampling and analysis program. Cells can be built above grade, below or both.

Once a cell has been prepared to accept waste materials, internal berming with clay provides for multiple subcells to segregate different types of waste. One subcell is used for pseudometal residues requiring pH control at or near neutrality. A second subcell houses heavy metal residues at an alkaline pH using carbonate buffering. The third and largest subcell is categorized "general" for the majority of wastes. A further control over placement involves the use of a three-dimensional grid system to preclude mixture of incompatible wastes and to provide for the use of eight different cover mixtures. The cover mixtures are made up of lime, soil, sand, fly ash, slag, clay and chemical additives such as ferrous sulfate and activated carbon, to impart dimensional stability and load-bearing strength as well as chemical buffering and waste attenuation.

After the wastes have been converted to a solid, landfillable form they are placed in the appropriate grid sector and covered daily. The placement is recorded on a daily grid chart which is a permanent record. Operation in this solid, dry mode permits vehicular traffic within the cell, minimizing the possibility of an accidental spill during handling and placement.

The internal leachate collected is monitored, then removed and treated through a physical/chemical aqueous treatment process. The impurities are returned to the cell in the form of a solidified sludge. A final dewatering of completed, capped cells results in a thereafter dry condition requiring only periodic inspection.

In conclusion, a practical system and an operational protocol have been defined to provide positive control over a broad spectrum of wastes. The proper use of chemical and physical pretreatments maximizes the applicability of secure landfilling. As alternative techniques for recovery or destruction of some components and volume reduction of others, it is probable that more residues will fall within the criteria for determining a hazardous waste, and hence secure landfilling will continue its importance, with a shift to more inorganic and fewer organic applications.

CHAPTER 5

SECURE LANDFILLS FOR CHEMICAL WASTE DISPOSAL

R. A. Johnson
Browning-Ferris Industries, Inc.
Houston, Texas

The high standard of living and the long life expectancy that we enjoy today is to a large extent the result of advances in the manufacture and use of synthetic chemicals. The quality of the food we eat, the medication we take, the clothes we wear and the automobiles we drive are all directly related to a vast array of chemicals and the efficiency and economy with which these chemicals are manufactured. As with most things in life, there are risks or trade-offs that are associated with such benefits. For example, when we get into an automobile or airplane, we are encountering the potential risk of a disabling or life-claiming accident. However, society has concluded that the benefits of these forms of transportation far outweigh the small risks that are involved. With the manufacture and use of some synthetic chemicals, we are faced with potential health risks if the materials are not properly used. In addition, we must accept the unavoidable generation of chemical waste by-products which must be disposed. Potential health risks are also encountered with some of these chemical waste materials if they are not handled properly in the treatment and disposal process.

The disposition of waste materials will involve one or more of the following:

1. treatment or processing for recovery of valuable components for recycling;
2. use as a fuel source for energy production;
3. treatment for discharge to the surface waters;
4. treatment and/or incineration for discharge to the atmosphere; and
5. treatment and/or disposal onto or into the land.

It is Browning-Ferris' belief that if a waste material can be processed for materials recovery or as a source of energy or if it can be treated and discharged to the air or surface waters, in most cases this will be done by the generator at the source of generation of the waste material. Economics will normally dictate this course of action. Generating plants are normally located to enable the use of proper treatment with permitted discharges to the surface water and air, or discharge to a municipal sewer system. If a chemical waste has valuable components that can be recovered economically or if the material can be used for fuel, the generator may normally best be suited to realize these benefits. This onsite disposal eliminates the high cost of transportation that is often associated with the transfer of the wastes for offsite disposal.

However, if a material cannot be processed for recovery or as a source of energy, and cannot be treated for discharge to the air or surface waters, it must be disposed into or on the land. There are no other alternatives. Browning-Ferris believes that proper land disposal can normally be best handled offsite and that this can be done best and more economically by the waste management industry. The reasons are again ones of economics. The hydrology and geology at many manufacturing locations are not conducive to proper land disposal. In addition, at many such locations sufficient land is simply not available.

The proper land disposal of chemical wastes starts in the laboratory. Starting with the chemical analysis of the waste available from the generator, additional information is generated as required to determine if the material can be safely disposed in a sanitary landfill. The decision tree shown in Figure 1 has been developed and is currently being used by Browning-Ferris for waste classification.

If the material "as received" is not suitable for sanitary landfill disposal, the next step is to determine if the material can be treated and stabilized chemically so that sanitary landfill disposal would be acceptable.

The following alternatives are available for the treatment of waste materials:

- chemical neutralization
- oxidation/reduction
- concentration (precipitation, emulsion breaking, etc.)
- chemical fixation
- solidification

The materials produced by the above treatment methods are then submitted to the same waste classification decision tree shown in Figure 1.

If this is unsuccessful, the next step is to determine the proper treatment and/or handling methods for disposal in a "secure landfill." A secure landfill is similar to a sanitary landfill except that much more care is taken to ensure that the wastes will never come into contact with the air, surface waters or groundwater. This isolation from air and water is achieved through the use of thick layers of essentially nonpermeable clay or a combination of clay and synthetic liners. See Figure 2 for the various landfill liner alternatives proposed in Section 3004 of the Resource Conservation and Recovery Act (RCRA). The wastes are surrounded on all sides by these impermeable barriers. When clay is used, the permeability of the clay must be less than 1×10^{-7} cm/sec.

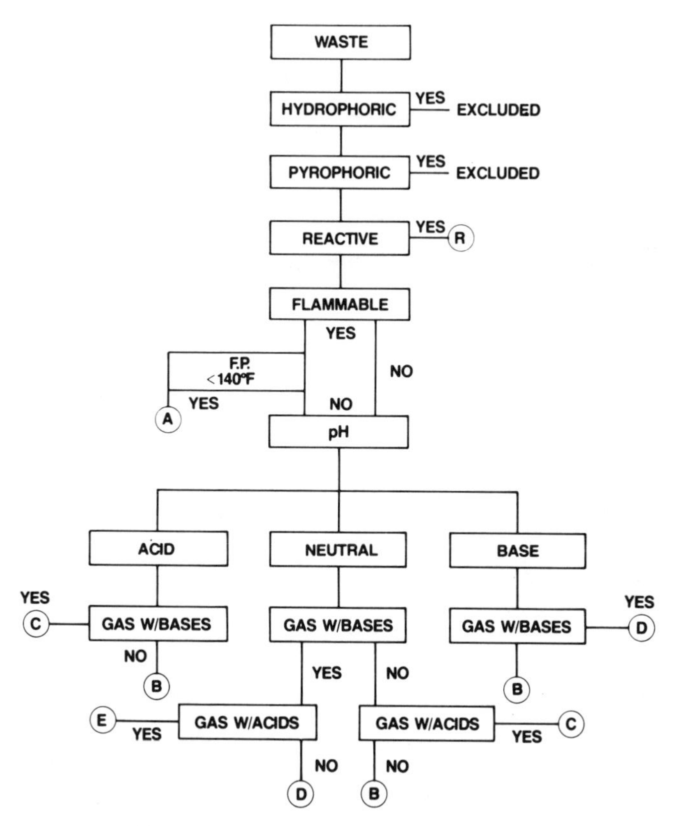

Figure 1. Waste compatibility decision tree.

What this means is that if you had a bowl made of this clay, and if you maintained one foot of liquid in the bowl, the liquid would only penetrate the clay at a rate of 0.12 inches per year. Thus, it would take ten years for the liquid to penetrate one foot of the clay *if* the free liquid level of one foot was maintained at a constant level. The outer walls and bottom of a secure landfill are always greater than three feet in thickness and are often thicker than ten feet. Utilization of these thick barrier walls, plus prohibiting the disposition of free liquid into a secure landfill, essentially eliminates the possibility of future leakage. When the landfill is completed, it is "capped" with a two foot layer of clay and six inches of soil for vegetation. The top is contoured to prevent ponding of rain water on the surface. This vegetated, contoured clay cover prevents storm water intrusion into the completed landfill, which could create a driving force for future chemical leakage.

As a further safeguard, many secure landfills also employ a liquid collection system above the bottom liner. Thus, if by some remote chance, free liquid

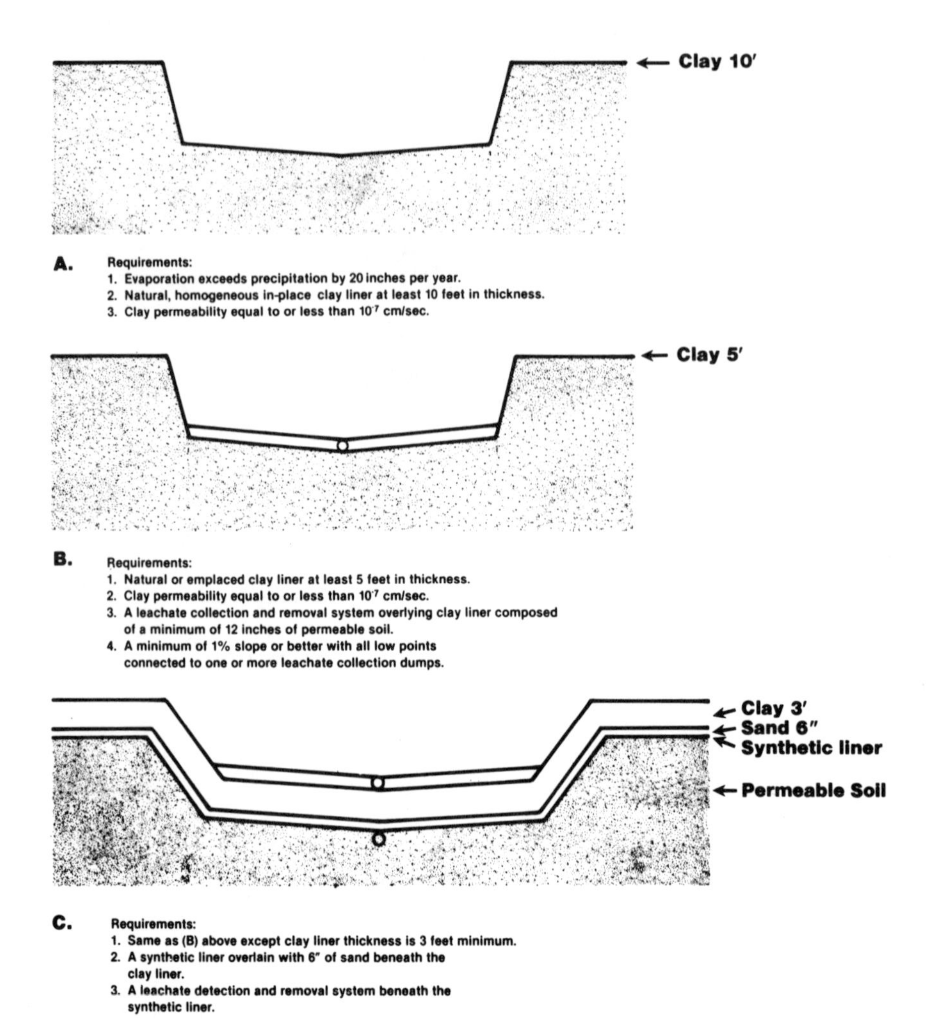

Figure 2. Secure landfill liner alternatives.

does somehow get to the bottom of the landfill, it can be recovered and removed so as not to impose any future leakage potential.

In the operation of a secure landfill, it is imperative to ensure that incompatible waste materials are not allowed to come in contact with one another. The waste compatibility decision tree shown in Figure 3 has been developed and is being used by Browning-Ferris to determine which wastes can be safely commingled. When it is necessary to separate waste materials, this is accomplished in the secure landfill by the construction of individual clay-lined cells.

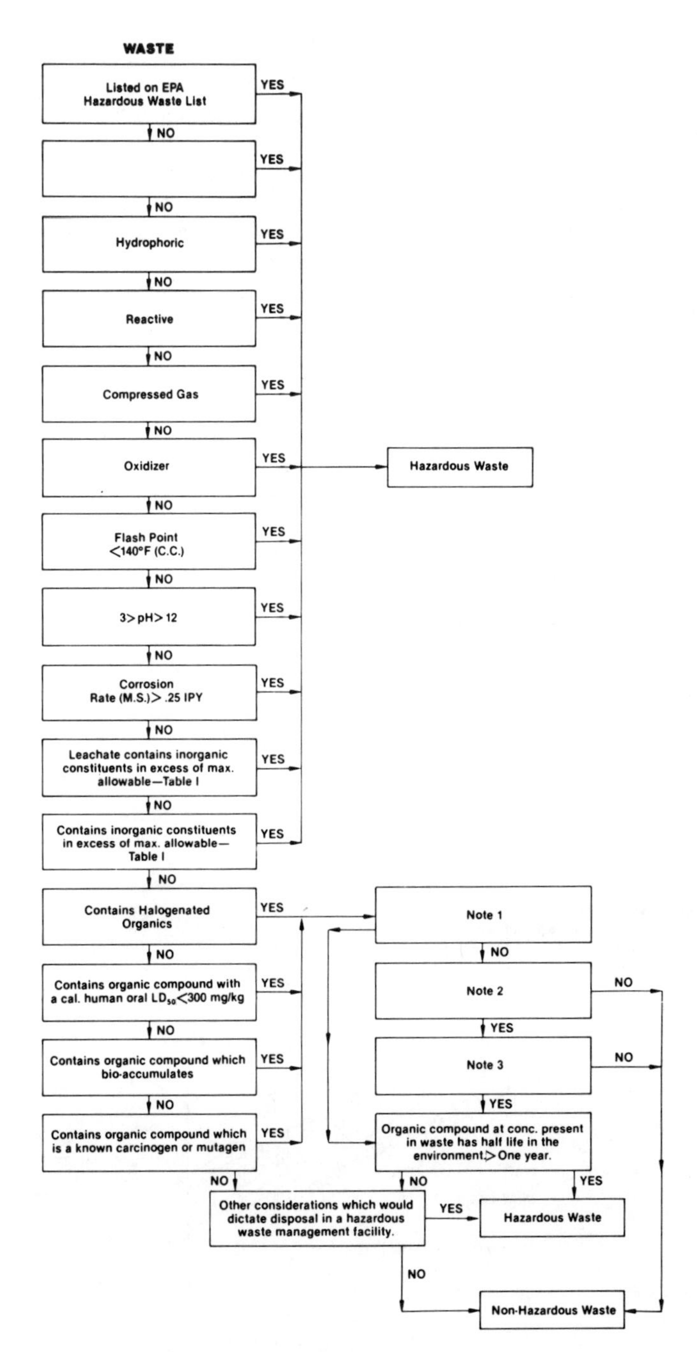

Figure 3. Waste classification decision tree.

Vertical walls 18-in. thick and 12-in. thick floors and caps of clay are used in constructing the cells to ensure that the waste will never become commingled. Figure 4 shows a typical cross section of a secure landfill with individual cells for segregation of waste streams.

In general, we prefer to dispose of only bulk solids or drummed solids into a secure landfill. With this type of operation, it is essentially impossible to have future problems created from free liquid within the landfill. However, at times for personnel protection or for other reasons, it is necessary to dispose of drums containing liquids. When this is necessary, sufficient dry chemical absorbent is placed around the drums to absorb the free liquid within the drums when the drums ultimately deteriorate. Thus, the future potential for free liquid within the landfill is eliminated.

When placing drummed materials into the landfill, an effort is made to provide the best possible environment around the drums for the future containment or degradation of the waste materials. For example, when drums of acidic materials are placed into the landfill, they are surrounded with a sufficient quantity of base, such as lime, to elevate the pH of the waste when the drums deteriorate.

Although the construction and operation of a secure landfill as described above is essentially fail-safe, additional safeguards are employed. Monitor wells are installed around the landfill to detect any leakage should it occur. In addition, records are maintained of the exact location and chemical composition of each waste that goes into the landfill. These records become part of

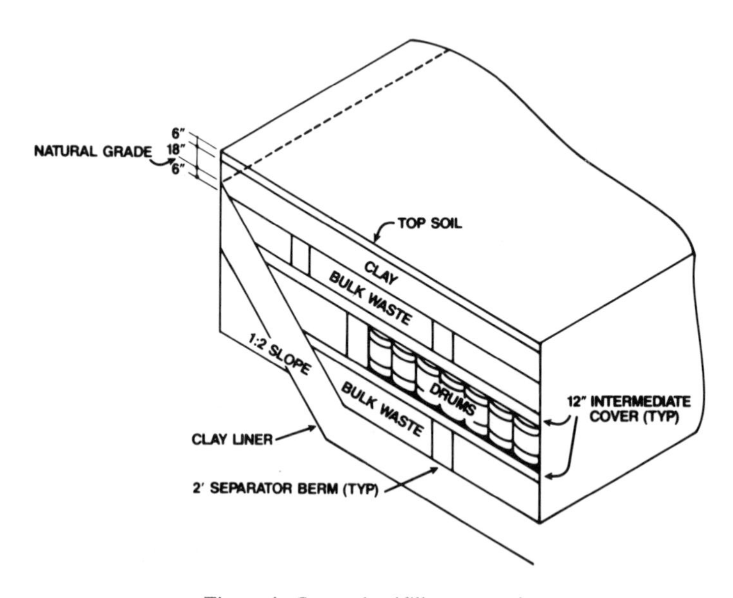

Figure 4. Secure landfill cross section.

the permanent deed on the property. Thus, any future problem that might occur can be more quickly traced to its source in the landfill and be resolved more quickly and efficiently. Also, retrieval of individual waste materials from the landfill might be possible if their future value would justify this course of action.

This method of land disposal provides total isolation from the environment of those worthless chemical wastes that would otherwise pose some degree of risks to the human health and environment. Thus, we are allowed to continue to enjoy a high standard of life with an absolute minimum of risk from exposure to the chemicals or chemical wastes that help to create this high standard.

CHAPTER 6

A SOLID FUTURE FOR SOLIDIFICATION/FIXATION PROCESSES

D. C. Christensen and W. Wakamiya
Battelle–Pacific Northwest Laboratory
Richland, Washington

INTRODUCTION

Air and water quality regulations in the U.S. have been evolving over the past two decades. These regulations dictate that a certain minimum level of environmental quality must be maintained to protect human health. As a result, suspended and dissolved pollutants must be removed from polluted effluents before discharge to the environment. The collected pollutants must also be disposed of in an environmentally safe manner. The need to treat and handle these solid wastes has resulted in the development of many new waste treatment processes. For gas treatment, various processes including baghouse, electrostatic precipitators, cyclone separators and liquid scrubbers have come into common use with a commensurate increase in fine particulate and wet sludge by-products. Water treatment processes such as chemical precipitation, evaporation, reverse osmosis, ion exchange and electrodialysis generate growing volumes of sludges and concentrated liquors.

In the past these residuals were relatively small and were generally discarded in landfills on company property or at community landfill sites. Since this could be accomplished simply and at low cost, little attention was directed to long-term management alternatives. In time, groundwater problems appeared in aquifers fed by contaminated leachate from these sites. Consequently, the uncontrolled disposal of potentially harmful substances in landfills has received intense scrutiny and will no longer be legal. This, in turn, has brought about interest in advanced waste disposal technology such as chemical solidification/fixation.

Until recently, treatment and disposal were simply links in the chain of events for management of process wastes. No thought was given to the ultimate fate of chemical wastes. Solidification/fixation processes were designed to address the long-term fate of wastes. As a result, these processes have come to be described as necessary steps for ultimate disposal.

"Ultimate disposal" means the final deposition of waste which, for either technical or economic reasons, may not be recycled or further reduced in volume by conventional treatment processes [1]. The overall mechanism by which ultimate disposal was achieved was through the conversion of the waste into a rigid solid which was structurally sound. In fact, in this context "solidification" is a term given to the conversion of liquids, sludges and solid debris into a solid, structurally sound material for fill, land reclamation and other useful purposes. [2].

The process must go one step further. It must not only be capable of solidifying the waste but also of binding the undesirable constituents in the waste stream to make them immobile in the environment. Undesirable constituents may include heavy metals and persistent toxic organics. Such materials are often too readily mobilized by leaching to be disposed of directly in landfills and must be "fixed" through physical/chemical methods. This constitutes fixation. Fixation itself may involve a variety of mechanisms ranging from an actual chemical bonding to physical adsorption or encapsulation.

Thus, a process aimed at the ultimate disposal of undesirable substances involves the reaction of a variety of chemical additives with the waste materials to form a chemically and structurally stable product. The process must make the waste essentially unleachable so that potentially harmful quantities of contaminants cannot migrate from the treated product. The process must also make the waste structurally stable so that during handling, disposal and long-term observation of the waste, it will not crumble and decompose, thus liberating its contaminants.

INORGANIC TREATMENT

As the variety and efficiency of the various pollution control processes has increased, large amounts of inorganic and mixed sludges have been generated which require treatment prior to land disposal. Recognizing this, innovative scientists and engineers have explored the possibility of reusing the waste itself as a means of safeguarding its disposal. A variety of cementitious methods have been developed to incorporate the dissolved constituents in the waste as a part of the rigid cement matrix.

The most effective systems involve the use of Portland cements, lime based mortars, lime-pozzolan cements such as lime fly ash and some mixed inorganic-organic materials. Portland cement, the most common material, is prepared by sintering fixed portions of calcium carbonate and aluminosilicate in a kiln at a very high temperature. During the firing, the materials combine to produce clinkers (small round lumps) which are then ground to become the cement powder. This powder is primarily made up of calcium silicates. The silicates give the cement its hydraulic character (or the property of hardening in the presence of water) [3].

The composition of commercial Portland cement is a mixed system of three oxides–calcium, silicon and aluminum–as well as various impurities such as iron oxide and lime. Hydration of silicate compounds is largely responsible for the hardening or "setting" of Portland cement/water mixtures. The hydration products form a colloidal calcium-silicate-hydrate gel. In the hardened cement this gel comprises about 70% by volume of the material and thus forms the main bonding between unreacted cement and other crystalline products of hydration [3]. It is these hydration products which are of most interest in treating persistent inorganic wastes.

In the presence of lime, many persistent heavy metals will combine to form colloidal metal hydroxides. The metal hydroxides then become an intimate member of the cement matrix. This stepwise process is demonstrated in Figures 1 through 3. In Figure 1 the circles represent unreacted particles of ground Portland cement. As water is added the cement particles begin to hydrate and form the calcium-silicate-hydrate gel as illustrated in Figure 2. Along with this gel formation comes the formation of various crystalline hydration products such as calcium hydroxide and various heavy metal hydroxides. These products form in the interstices of the cement matrix. During the final stages of hydration the gel swells to the point where particle overlap occurs and silica fibrils develop. These finger-like fibrils are represented in Figure 3. All of the hydration by-product crystals have grown to their maximum size and are either overlapped by fibrils or have grown into the particle gel itself. The interlocking of the fibrils and the formation of various hydration products binds the cement and other components of the mix into a rigid mass.

Unreactive materials that are blended with the cement prior to gel formation and setting can become encased in the solid matrix. In many cases these materials can provide cohesive strength to the final product. When using Portland cement as a building material, various sizes of unreactive aggregate are added to the cement for the purpose of adding strength. This same process occurs when adding mixed sludge wastes to the cement. Unreactive, consolidated portions of the sludge can become encased in the matrix and can impart additional structural quality to the rigid mass. These materials are not, in fact, chemically bound to the solid but their mobility is restricted because of physical entrapment.

Portland cements have been used to immobilize radioactive and military wastes [4], heavy metal industry waste sludges from plating operations, toxic industrial wastes such as arsenic-bearing sludges, and for stabilizing silts and solids for land reclamation.

A problem encountered in achieving stabilization is that each waste has its own particular set of chemical requirements. Slight alterations in chemical additives to the unreacted cement may be required to achieve the most stable product. Often, in large chemical complexes, the wastes from a number of different processes are blended prior to treatment. This can create a very

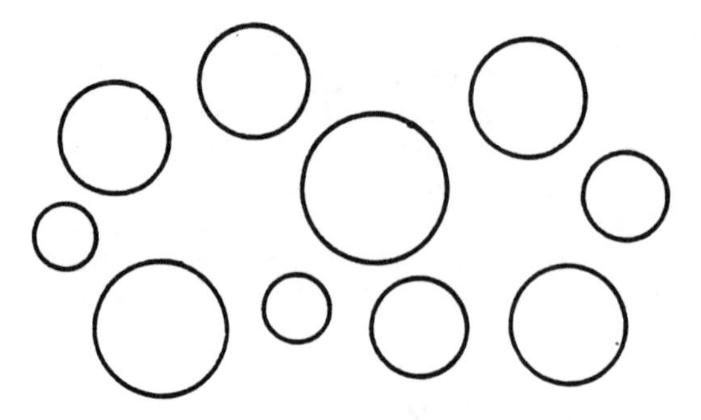

Figure 1. Representation of the grains of Portland cement before hydration.

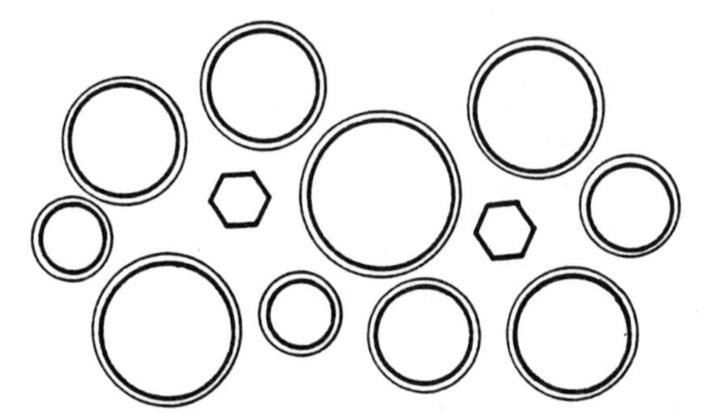

Figure 2. Silicate gel begins to form around each grain as the calcium silicate dissolves. Small crystals are various hydration by-products such as lime and metal hydroxide precipitates.

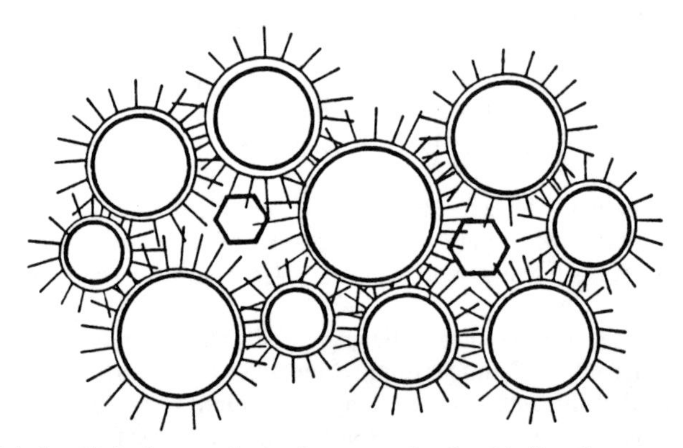

Figure 3. Tubular fibrils form as hydration proceeds, thus binding the cement and other hydration products into the hardening mass.

difficult problem when designing a specific chemical mix for treatment. Avenues for approaching this problem must be addressed on a case-by-case basis.

ORGANIC PROCESSES

One of the most difficult problems encountered in immobilizing a waste is that of persistent organic contaminants such as polychlorinated and polybrominated biphenyls (PCB and PBB), Kepone®, Mirex, TCAB and dioxin (TCDD). The Clean Water Act, as well as the Resource Conservation and Recovery Act (RCRA) guidelines, have addressed toxicity problems with the need for removal of certain persistent organic contaminants from the environment. Although many of these organics are only sparingly soluble in water, they nevertheless still impose a large toxic impact through bioaccumulation. They are also generally unreactive with many common chemical reagents and their solubility may be affected by alterations in the chemical environment. This does not imply that any chemical alteration of the organic is taking place. When attempting to fix this type of contaminant in a cement matrix, little, if any, chemical bonding takes place. The only effective immobilization mechanism is that of encapsulation. This is often insufficient because the encapsulation mechanism can be a weak arrangement which allows chainlike organic molecules to move through the interstices of the silicate gel and fibrils when only a small driving force is applied.

Until recently there has not been a great demand for solidifying predominantly organic-contaminated industrial wastes. However, there is a large variety of possible treatment chemicals which could be beneficially applied for solidification and immobilization of organic contaminants. At the present, the most available materials include asphalts, tar, polyolefins and various epoxies. These materials are either synthetic or a result of processing chemical feedstocks, and are generally more expensive than Portland cement. Therefore, the cost of a general application of these materials to a large toxic organic sludge problem must be weighed against the cost for alternative disposal (such as incineration). There have been isolated instances where small volumes of hazardous organics have been treated with tars and asphalts. In these instances the unit cost was found to be high, but the total capital outlay was quite small.

There are a variety of problems encountered in treating organic sludges. Probably the most persistent problem is that of treating hydrophobic materials in a water-based sludge. Surface tension effects themselves can become driving forces in mobilizing trace organics trapped within a solidified matrix. The effects of pH can also increase the mobility or leachability of the trace organic. Microorganisms can attack both the organic fixation agent and the organic contaminant and cause a rapid breakdown of the solid matrix or an alteration of the toxic substance into a more or less mobile molecule. Last but not least, the effects of temperature and ultraviolet radiation can greatly affect the leachability of the contaminant from the solidified matrix. Temperature can cause breakdown of the fixation agent, stimulate microorganism growth or increase the toxic constituent mobility through molecule excitation. Ultraviolet radiation can facilitate the breakdown of the fixation agent.

Since incentives for treating waste organic sludges are relatively recent, there has been little work in designing and evaluating various organic fixation processes. Previous work has involved the use of materials employed for other purposes but which form rigid solids. There has been no emphasis on designing specific materials or additives for the organic solidification/fixation problem. Consequently, the existing processes are expensive for waste treatment. If efforts are to be made to reduce the cost of treating and disposing of contaminated waste organic sludges, more emphasis is going to have to be turned to specific chemical design and selection.

SPECIFIC CASE RESULTS

Test results from the use of inorganic solidification agents have reported successful use of silicate-based agents on various refinery wastes, electronic industry wastes, power plant fly ash and others. Sealosafe, whose process was developed in England, reported the successful use of their silicate-based agent on a synthesized industrial waste having over 400,000 ppm total dissolved solids (TDS) of heavy metal contaminants. The resulting leach water had only 2 ppm TDS of heavy metals [5]. The waste contained hexavalent chromium, copper, nickel, zinc, lead, cadmium, manganese, sulfide, tin and arsenic.

Sealosafe has also reported, in their information bulletin 74-28, the effective use of their fixation agent in treating other industrial wastes. These results are summarized in Table I [5].

The Chemfix process [1] has been used on many of the wastes as reported by Sealosafe. It too is a silicate-based solidification agent which forms long chain inorganic polymers as it hardens. The results of the use of Chemfix on an electronic company's heavy metal-bearing waste can be found in Table II [1]. Similar results have been reported for steel plant wastes, refinery wastes (heavy metal wastes only) and power plant wastes [1].

These results are only a few of the many reported by the various companies involved in chemical solidification technology. They report very successful use of silicate based agents in immobilizing heavy metal sludges. Little has been reported on the use of fixation agents on organically contaminated sludges.

In 1977 EPA initiated steps toward clean-up and detoxification of a major portion of the James River, Virginia, which had been contaminated by the pesticide Kepone [6]. One method investigated for Kepone immobilization in the sediments was chemical fixation. A number of companies active in the chemical fixation business were contacted and asked to participate in a screening test of treatment agents. A list of companies contacted can be found in Appendix A.

The materials tested included silicate-based agents (Portland cement derivatives), organic-based polymers, a sulfur-silicate blend, a gypsum-based material

Table I. Leaching Data of Wastes Treated by Sealosafe

Type of Waste	Concentration of Pollutants in Waste (ppm)	Concentration of Pollutant in Leachate (ppm)
Power Generation Ash	28,000 V	0.16 as V
	2,000 Ni	0.1 as Ni
Metal Finishing Chrome Sludge	87,000 Cr	0.15 as Cr
	5,200 Cu	0.05 as Cu
	22,000 Pb	0.1 as Pb
Petrochemical Catalyst	7,800 Co	0.05 as Co
	16,800 Fe	0.1 as Fe
Spent Gas Absorbant	170,000 Z	0.15 as Zn
	1,400 Sulfide	0.1 as Sulfide
Rubber Industry	16,000 Sb	0.5 ppm as Sb
Tin Production Waste	16,200 As	0.25 as As
	4,000 Pb	0.1 as Pb
Sodium Slag	105,000 Sulfide	0.1 as Sulfide
Acid Neutralization	560 Hg	0.005 as Hg

Table II. Leaching Data of Electronics Industry Waste Treated by Chemfix

Constituent	Company Analysis (10/19/71)	Chemfix Analysis (1/17/72)	After 3 Liters of Leach Water
Cr	100– 1,000	134	0.05
Fe	5,000–50,000	106	0.025
Zn	100– 1,000	137	0.025
Ni	5– 100	32.3	0.10
Cd	10– 100	19.6	0.10
Mg	500– 5,000	232	0.05
Cu	5– 100	5.1	0.05
Al	1,000–10,000	[a]	0.10
Pb	5– 50	[a]	0.10
Mn	500– 5,000	[a]	0.10

[a]Not analyzed.

and molten sulfur. Not all companies contacted chose to participate in the testing program. Those companies who did generally sent more than one chemical mix. The results of some of the more successful materials are summarized in Table III [7]. The procedure used for leach testing can be seen in Appendix B.

Table III. Kepone Concentrations in Leachate Solutions ($\mu g/1$, ppb)

Fixation Type	Time (days)								Composite of Leachate	Calculated Composite
	0.04	0.17	1	7	14	28	56	84		
Silicate Base										
1	0.07	0.08	0.094	0.166	0.524	0.30		0.26	0.17	0.21 ± 0.16
2	0.05	0.05	0.111	0.157	0.306	0.26		0.51	0.26	0.21 ± 0.17
3	0.095	0.068	0.059	0.15		<0.21	0.31	0.77	0.77	0.24 ± 0.25
4	0.21	0.24	0.21	0.16		0.096	0.14	0.81	0.83	0.27 ± 0.24
5	0.046	0.088	0.068	0.67		0.033	0.11	0.15	0.24	0.17 ± 0.23
Organic Base										
6	0.042	<0.075	0.021		<0.055	0.034	0.057	0.021	<0.049	0.043 ± 0.02
7	0.086	0.044	0.010	0.053	0.096	0.21	0.28	0.083	0.074	0.11 ± 0.09
Sulfur Base										
8	0.013	0.012	0.017	0.010	0.05	0.029	0.032	0.15	0.17	0.039 ± 0.046
9	0.5	0.22	0.095	0.20		0.28	0.31	0.29	0.45	0.27 ± 0.12
Blank										
10	<0.066	<0.066	0.076	0.058	0.050	0.22		1.04	0.10	0.225 ± 0.36
11	0.117	0.04	0.104	0.081	0.11	2.30		0.14	0.26	0.44 ± 0.28
										0.10 ± 0.035[a]

[a]Value recalculated excluding 28 day sample.

In general the silicate-based agents did not effectively reduce the leachate Kepone levels below those of the unfixed sediments. Many times they increased the Kepone concentration. This is believed to be the result of the reliance on Portland cement and the associated high pH of the reagents. Since Kepone solubility is greatly increased at higher pH levels, the silicate-based agents release Kepone otherwise bound to the sediments.

Kepone Solubility (μg/l)	pH
0.03	7.0
0.035	8.0
0.050	9.0
0.090	10.0
0.30	11.0
3.00	12.00

Of the two organic polymers tested, both were quite effective in reducing the concentration of Kepone in the leachate by an order of magnitude. However, the polymer sealant relies on a film-like coating. When this was broken for the elutriate test, the Kepone became more readily available. Hence, the sealant is more appropriate as a soil amendment while the polymer grout would function well for spoil stabilization.

Molten sulfur was found effective in reducing Kepone availability below that of untreated sediments. The sulfur-cement proprietary blend performed at a level intermediate between the molten sulfur and the silicate-based materials. The gypsum-based material was ineffective. As a result of these evaluations, it was determined that the organic grout and the molten sulfur hold the greatest promise as stabilization agents for Kepone-contaminated spoils.

Concentration vs time plots for leachate samples of five of the materials have been plotted in Figure 4. The center line represents the leach concentration from the untreated blank material. The two upper lines represent the leach concentrations from the silicate-based solidification agents while the two lower lines represent the leach concentration from the organic grout in molten sulfur agents. Clearly the organic epoxy grout and the molten sulfur are the most effective in treating the Kepone-contaminated sediment. A more extensive discussion of the approach taken on this work and the results derived from the work can be found in McNeese et al. [8].

In light of the initial screening tests, additional laboratory studies, sponsored by Allied Chemical Corporation, were performed to evaluate the effectiveness of the epoxy grout in treating highly contaminated industrial sludges and solid wastes.

The purpose of the additional work was to determine the effectiveness of the epoxy grout sealant in fixing a co-contaminated (organic-inorganic) industrial waste, as well as treating a waste of high contaminant concentration. The organic contaminant was Kepone and the inorganic contaminant was arsenic.

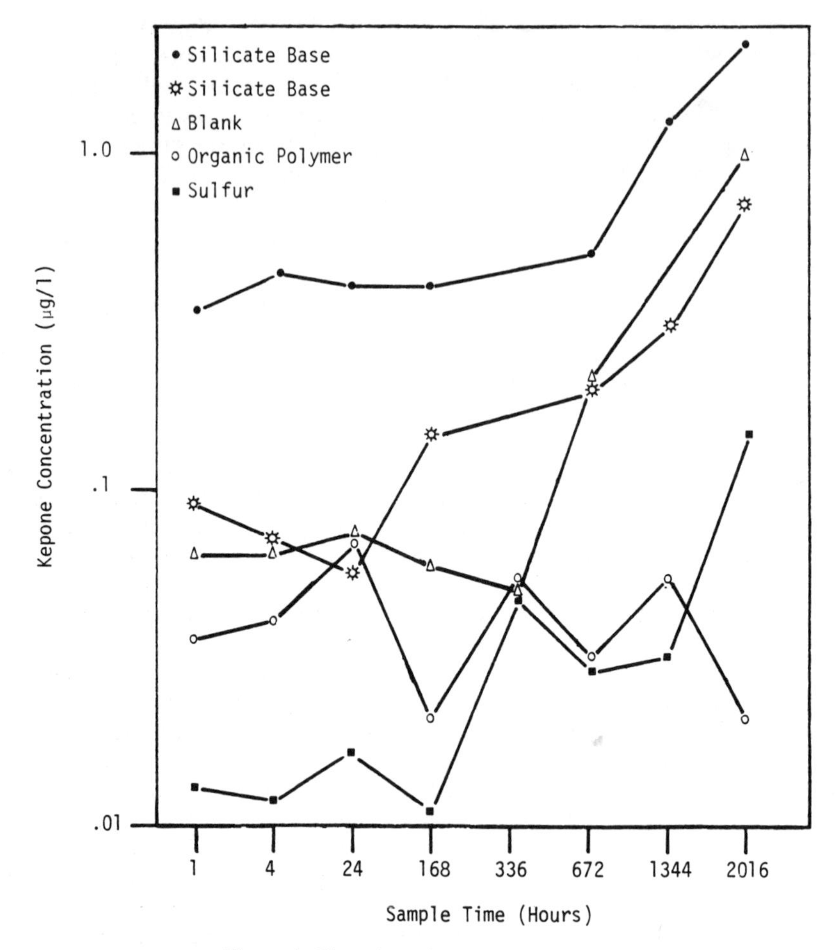

Figure 4. Time-dependent leachate data.

Three discrete sludge samples were tested. Each sample was blended and prepared for solidification. The samples had the following arsenic and Kepone* concentrations:

	Arsenic (mg/g)	Kepone (mg/g)
Sample A	56.3	5.2
Sample B	20.7	30.0
Sample C	19.6	42.4

*All Kepone concentrations are the totals of Kepone and Kepone degradation products.

The procedure used for leach testing can be seen in Appendix B.

Over an 8-week test period leachate evaluations of fixed samples showed 74.3% retention of arsenic and 99.5% retention of Kepone in Sample A; 96.6% retention of arsenic and 98.6% retention of Kepone in Sample B; and 99.6% retention of arsenic and 99.7% retention of Kepone in Sample C. Results for the leach test are summarized in Table II.

The leach test analyses were made at one, two, four, six and eight weeks. The control samples are duplicates of unfixed sludge exposed to the same conditions as the fixed samples. The results suggest what the long-term impact might be if the solidified sludge remained undisturbed as moisture percolates through it.

As can be seen in Table IV, Sample A had the lowest arsenic retention percentage of the three samples, but it also had the lowest "control" concentrations of the three samples. The resulting "fixed" sample arsenic concentrations were very close to the drinking water standards of 0.05 ppm. The leachate concentration decreased over the 8-week period from 0.117 ppm to 0.030 ppm. At the two-week sampling interval, the fixed leachate concentration of 0.044 was already below the drinking water standard.

The fixation of Sample C was very successful with respect to arsenic retention. The concentration of arsenic in the control blanks is approximately

Table IV. Leachate Test Results for Arsenic and Kepone Concentrations (mg/l)

Sample	1	2	4	6	8	Composite	Percent of Control
Arsenic[a]							
Control	0.38	0.30	0.18	0.16	0.17	0.25	
Fixed	0.117	0.044	0.056	0.040	0.030	0.065	74.3
Kepone[a]							
Control	193.0	50.1	21.7	6.6	2.5	50.4	
Fixed	0.281	0.177	0.310	0.234	0.235	0.245	99.5
Arsenic[b]							
Control	99.0	55.9	27.5	15.5	10.3	39.3	
Fixed	0.80	0.99	1.92	1.72	1.85	1.35	96.6
Kepone[b]							
Control	0.334	0.494	0.360	0.257	0.157	0.280	
Fixed	0.002	0.011	0.004	0.009	0.002	0.004	98.6
Arsenic[c]							
Control	2.52	1.61	2.85	2.19	1.67	2.12	
Fixed	0.006	0.004	0.013	0.010	0.009	0.008	99.6
Kepone[c]							
Control	2.18	---	65.3	---	2.84	32.3	
Fixed	0.024	---	0.109	---	0.136	0.084	99.7

[a]Sample A.
[b]Sample B.
[c]Sample C.

three orders of magnitude higher than the arsenic concentration of fixed samples. The arsenic concentration in the fixed samples ranged from 0.004 ppm to 0.013 ppm. In all cases the arsenic level is well below the acceptable drinking water concentration.

Concentrations of Kepone in the leachate from all three fixed samples were reduced by 2-3 orders of magnitude from the untreated sludge. None of the leachate results for Kepone could meet the proposed ambient water criteria of 0.008 ppb.

Based on the solidification testing, for every ton of sludge treated, 1 ton of aggregate and 27 gal of epoxy will be required. The cost of the aggregate is approximately $200/ton and the cost of epoxy is $14.41/gal. Therefore, the chemical costs to treat each ton of sludge will be $590 or 30¢/lb of sludge. In addition to these, other costs involving transportation, mixing, labor, equipment, leachate and residual water treatment, and long-term management will have to be added. The cost of epoxy grout treatment will likely restrict its application to only small-volume or highly concentrated wastes.

Further experimentation on the use of molten sulfur as a fixation material has also begun. Initial tests involving 20:80 mixes of molten sulfur and fly ash have been fairly inconclusive. As with the James River Kepone results, it is expected that a 50:50 mix of sulfur and waste will provide adequate fixation. At a 50:50 mix, a ton of fly ash will require a ton of sulfur. At $40 per ton for sulfur, this makes the chemical cost for sulfur fixation at $40/ton of waste, or 2¢/lb.

CONCLUSIONS

RCRA and the Toxic Substance Control Act (TSCA) will greatly affect the way industries must handle and dispose of their wastes. It appears that technology is available to address the ultimate disposal requirements for toxic metal and trace metal wastes. The technology for wastes containing persistent organic contaminants is not as readily available. Current materials used for fixation are either inadequate for organic control or not designed specifically for the ultimate disposal of persistent organics.

Two materials pointed out in this paper appear to be likely candidates for additional investigations: molten sulfur and an epoxy grout material. The epoxy grout has been tested further for use as a fixation agent on both a co-contaminated (organic-inorganic) waste and a waste of high contaminant concentration. It has proved to be very successful for treating both cases.

Since the organic grout is not designed specifically for Kepone fixation, or any other contaminant fixation for that matter, it likely does not provide an optimal fixation capability. However, it does confirm that fixation chemicals are available in a raw form which are capable of immobilizing organic contaminants.

At this time the major problems encountered in using these chemicals are the cost of treatment and the availability of a variety of epoxy materials suitable for treatment. At a chemical cost of 30¢/lb of sludge treated, the process is only applicable to small-volume wastes and wastes of high contaminant concentration. It likely could not be applied to large volumes of waste. At a chemical cost of 2¢/lb treated, molten sulfur can be applied over a large range of waste volume but it is still an expensive approach to chemical fixation.

Efforts must be turned toward the synthesis of specific fixation chemicals capable of immobilizing hazardous organic wastes. Lower treatment costs may be obtained through lower epoxy mixing ratios or the selection of an epoxy capable of selectively immobilizing the contaminant. Further investigations and material development must be performed.

Bench-scale testing of the solidified waste must take place before applying the solidification/fixation agent to the general waste before disposal. Bench-scale testing of fixed samples will give information concerning expected leach rates, optimum solids to agent application ratios, and durability of the solidified sample. As a final requirement, some form of field demonstration will have to be performed to test the long-term durability and acceptability of the treatment approach.

REFERENCES

1. Conner, J. R. "Ultimate Disposal of Liquid Wastes by Chemical Fixation," in *29th Annual Purdue Industrial Waste Conference Proceedings*, West Lafayette, IN: Purdue University.
2. Krofchak, D., and J. N. Stone. *Science and Engineering for Pollution-Free Systems* (Ann Arbor, MI: Ann Arbor Science Publishers, Inc., 1975).
3. Double, D. D. and A. Hellwell. "The Solidification of Cement," *Scientific Am.* (July 1977), pp. 82–90.
4. Lokkan, R. O. "A Review of Radioactive Waste Immobilization in Concrete," prepared for U.S. Department of Energy, PNL-2654, Battelle, Pacific Northwest Laboratory (June 1978).
5. Chappell, C. L. "Disposal Technology for Hazardous Wastes," *Surveyor November 2, 1973* (Thornton Heath, Surrey, England: George Rose Printers).
6. Chigges, J. A. "Kepone Sediment Monitoring in the James River, Virginia During 1976," State Water Control Board Memorandum to File (July 6, 1977).
7. Dawson, G. W. et al. "The Feasibility of Mitigating Contamination in the James River Basin," in *Mitigation Feasibility for the Kepone Contaminated Hopewell/James River Area–Appendix A*, U.S. EPA (1978).
8. McNeese, J.A., G. W. Dawson and D. C. Christensen. "Laboratory Studies of Kepone®-Contaminated Sediments," in *Toxic and Hazardous Waste Disposal, Vol. 2*, R. B. Pojasek, Ed. (Ann Arbor, MI: Ann Arbor Science Publishers, Inc., 1979), pp. 217–228.

APPENDIX A

CANDIDATE FOR FIXATION AGENT TESTING ON KEPONE CONTAMINATED SEDIMENTS

Company	Address	Response
Chemfix-National Environmental Controls, Inc.	Metairie, LA	Fixed samples at company
TJK Ind.-U.S. Representatives for Takenaka	North Hollywood, CA	Fixed samples at company and sent chemicals for in-house fixation
IU Conversion System	Plymouth Meeting, PA	No response
Protection Packaging	Louisville, KY	Not developed enough for testing on sediments
John Sexton Landfill Contractors	Oakbrook, IL	No response
Werner and Pfleiderer, Corp.	Waldwick, NJ	No response
Wehran Engineering Corp.	Middletown, NY	No response
Ontario Liquid Waste Disposal Systems	Cambridge, Ontario	Fixed samples at company
TRW, Inc.	Redondo Beach, CA	Not developed enough for testing on sediments
Manchek Colorado	Santa Barbara, CA	Fixed samples at company
Key Chemicals	Philadelphia, PA	Sent samples of sediment, fixed at company
Dowell Division, Dow Chemical	Tulsa, OK	Sent chemicals for fixation and testing
Hallemite Division, Sterling Drug Co.	Montvale, NJ	Sent chemicals for fixation and testing
Randustrial Corp.	Cleveland, OH	Sent chemicals for fixation and testing

Tidewater Terminal Co.	Pasco, WA	Sent asphalt sample for fixation and testing
Surcoat (Chevron)	San Francisco, CA	No response

APPENDIX B

LEACHATE TEST PROCEDURE

1. All samples were held at 4°C before testing.
2. A 50-ml portion of each sample was split in half. One half is solidified with the fixation chemicals and the other half is used directly as a control blank.
3. Each of the split portions was placed in sealed leaching bottles and 500 ml of distilled water at pH 4.5 was added.
4. At each sampling interval the water was removed from the vessel and a new aliquot of pH 4.5 water was added.
5. Removed aliquots were split. Half was analyzed for the desired contaminant while the other half was collected as a composite sample. At the end of the testing period the composite samples were analyzed for total contaminant concentration.

CHAPTER 7

LEACHATE TESTING OF HAZARDOUS CHEMICALS FROM STABILIZED AUTOMOTIVE WASTES

N. I. McClelland, H. B. Maring, T. I. McGowan and G. E. Bellen
National Sanitation Foundation
Ann Arbor, Michigan

INTRODUCTION

The effectiveness of a service developed in England for treating hazardous industrial wastes and making them suitable for land disposal was demonstrated in a special study with wastes from a major automobile manufacturing facility in Michigan. The service, known commercially as SEALOSAFE(SM), is one which includes a process reported to stabilize liquid or solid industrial wastes (including sludges) to easily handled, nonhazardous materials by converting them to stable, solid polymers, STABLEX(TM), with the consistency of cement. According to the claims, toxic substances in the waste are fixed within the structure of the polymer. When the processed (stabilized) wastes are exposed to groundwater, toxic materials are not expected to leach to the environment.

In considering the feasibility and desirability of locating a STABLEX plant(s) in Michigan, the Michigan Department of Natural Resources (DNR) requested that the processor arrange for an independent study in which selected samples of hazardous wastes would be stabilized and their leaching characteristics demonstrated. The study was sponsored by Stablex-Ruetter, Inc., a subsidiary of Stablex Corporation, Radnor, PA, and undertaken by the National Sanitation Foundation (NSF), Ann Arbor, MI.

NSF

The National Sanitation Foundation is an independent, nonprofit organization chartered under the laws of Michigan. Principal offices and laboratories are located in Ann Arbor, Michigan, with regional offices across the U.S. and in Europe. Since 1944 its objectives have been to serve environmental and public health needs through service, research and education. Its more than 50 standards—and testing to assure conformance with these standards—are used as code, regulation or policy by regulatory authorities around the world.

The NSF standards development process requires that all areas of concern participate—regulatory agencies, manufacturers and users. The standard, when adopted, represents a consensus of these viewpoints, assuring that its requirements are both desirable and achievable. The NSF role is one of neutrality, both in developing the standard and in providing the objective testing services which follow. Ongoing testing programs include analysis of water for chemical and organoleptic characteristics leached from plastic piping system components, currently the largest testing program, and testing of water and wastewater treatment equipment for home, commercial and marine applications. As a U.S. Coast Guard and Canadian Environmental Protection Services recognized testing facility, marine sanitation device testing by NSF professionals may occur at a test site near Ann Arbor (Chelsea, MI), aboard ships at sea or in manufacturing facilities around the world.

With objectivity and analytical reliability clearly established, it is not uncommon for NSF to be asked to undertake a special study like leachate testing of hazardous chemicals from raw and processed industrial wastes.

PROTOCOL

The study was accomplished in two phases: (1) qualitative and quantitative analyses of raw wastes, and (2) analysis of processed wastes.

Five samples of automotive wastes—three from plating operations, one from paint priming and one waste treatment plant sludge (WTS) were selected for the study. The plating wastes were principally nickel, chromium and copper; the paint priming waste was principally zinc phosphate ($ZnPO_4$). WTS was a sample of sludge from a treatment plant receiving industrial wastes only (no sanitary wastes) from an automobile assembly plant. At this facility, 98% of the raw waste is attributable to painting and washing (metals preparation). A physical-chemical treatment process produces a sludge which, like the other wastes in this study, must be disposed of in accordance with regulatory requirements.

Phase I

Samples of the wastes were collected by the automobile manufacturer and delivered to NSF by NSF staff. The manufacturer, DNR and Stablex-Ruetter developed a list of "pollutants of interest" associated with the wastes to be studied. The list, shown in Table I, was reviewed with DNR and adopted for Phase I testing.

Phase II

Phase I data were used in developing the protocol for Phase II. Selection criteria included:

Table I. Parameters Measured in Phase I

Metals		Organics	Nutrients	Others
Aluminum (Al)	Manganese (Mn)	Total Organic Carbon (TOC)	Ammonia (NH_3)	Chloride (Cl)
Antimony (Sb)	Mercury (Hg)	Chemical Oxygen Demand (COD)	Nitrate (NO_3)	Cyanide (CN)
Arsenic (As)	Nickel (Ni)	Total Ash	Phosphorus (P)	Fluoride (F)
Barium (Ba)	Selenium (Se)			Sulfate (SO_4)
Beryllium (Be)	Silver (Ag)			Sulfide (S^{2-})
Cadmium (Cd)	Thallium (Tl)			Sulfite (SO_3)
Chromium (Cr, total)	Tin (Sn)			Thiocyanate (SCN)
Copper (Cu)	Vanadium (V)			Conductivity
Iron (Fe)	Zinc (Zn)			pH
Lead (Pb)				

1. Phase I parameters referenced in the National Interim Primary Drinking Water Regulations [1], published by the U.S. Environmental Protection Agency (EPA) (proposed regulations for hazardous waste were not published when Phase II was implemented);
2. metals for which the hydroxides are amphoteric (because the polymer buffers exposure water to pH $\geq$7.0, e.g., Al and Zn);
3. pollutants present at very high levels (e.g., Fe, F and P, measured in raw $ZnPO_4$ waste at levels of 3250, 3900 and 13,520 mg/l, respectively); and
4. parameters used to monitor the progress of Phase II testing (conductivity and pH).

Although the STABLEX formulation used in this study was not developed specifically to treat organics present in the wastes, total organic carbon (TOC) and chemical oxygen demand (COD) were included in Phase II to measure the leaching of gross organics and to demonstrate whether or not organics would affect stabilization of the inorganic constituents. Phase II parameters are listed in Table II.

After Phase I, NSF staff delivered samples of the raw wastes to Stablex-Ruetter, Inc., Camden, NJ, and observed the STABLEX application. Processing was accomplished in a bench-scale system by a Stablex-Ruetter chemist under the direction of Dr. C. Chappell, inventor of STABLEX. Up to ten ingredients were mixed in developing the polymers for these wastes.

METHODOLOGY

Two types of exposure testing* were included in Phase II, an equilibrium leaching test (ELT) described by Taub [2], and an American Society for Testing and Materials (ASTM) proposed method for leachate testing [3].

*The EPA method was not published at the time this study was undertaken. Subsequent attempts to develop comparative data using this method were inconclusive because very little of any of the samples were available at the end of the study.

Table II. Parameters Tested in Phase II

Pollutant	Processed Wastes				
	Ni	Cr	Cu	$ZnPO_4$[a]	WTS[b]
Al	X		X		X
As					X
Cd				X	
Cr (total)		X		X	X
Cu	X		X		
Fe				X	X
Pb				X	X
Mn				X	X
Hg				X	
Ni	X	X		X	
Se			X		
Zn		X		X	X
TOC	X	X	X	X	X
COD	X	X	X	X	X
Cl			X		
CN	X		X		X
F				X	
SO_4		X			
NO_3 (as N)				X	
P (as P)				X	
Cond., μmhos/cm	X	X	X	X	X
pH	X	X	X	X	X

[a]Applicable to both procedures (equilibrium leach test and ASTM Proposed Procedure).
[b]Waste treatment plant sludge.

Equilibrium Leaching Test

The ELT provided greatly accelerated exposure conditions when compared with those expected with land disposal of the processed wastes. After 14 days of curing (for the first day after processing, samples were allowed to air dry; for the next 13 days, they were sealed in airtight containers to simulate landfill disposal conditions), STABLEX samples were ground to a fine powder with mortar and pestle. Four consecutive exposures to carbon dioxide-saturated distilled water followed in covered beakers, each at a ratio of one part sample (150 g) to ten parts water by weight. The pH of the exposure water was 4.5 to provide an aggressive testing environment. A control (CO_2-saturated distilled water at pH 4.5) was included with each set of exposures.

All exposures were made at room temperature (approximately 25°C). Samples were exposed for periods of 1, 7, 14 and 28 days. During each 24-hr period, samples were mixed vigorously by magnetic stirrer for 1 hour. Conductivity and pH were measured immediately after each mixing period. After each period of exposure (1, 7, 14 and 28 days), the water was decanted, passed through a coarse filter (Whatman No. 42 "ashless"), then through a 0.45-μ filter, and analyzed for the pollutants listed in Table II.

All analyses (Phase I and II) except TOC were performed by NSF. TOC was done at the University of Michigan by university personnel. Metals were measured by atomic absorption spectrophotometry; nutrients, sulfite and thiocyanate were measured colorimetrically; COD, by wet chemistry (dichromate reflex); sulfate and total ash, gravimetrically; and chloride, cyanide, fluoride and sulfide, electrometrically (ion-selective electrodes). Details associated with these methods are included in Appendix A.

ASTM Proposed Method

This method was used for leach testing of raw and unpulverized processed zinc phosphate wastes to compare leaching from a raw and STABLEX processed waste, and to compare results from ELT with those from a proposed standard procedure.

A control (CO_2-saturated distilled water) was included with each exposure. A 393-g sample of processed waste was used for this test; therefore, 393 g of raw sludge were also used. Water (CO_2-saturated distilled) at four times the sample weight was added to each sample. The sample-water mixtures were placed in covered beakers and mixed in a water bath shaker (sixty 1-in. strokes/min) at room temperature for 48 hr. At the end of this period, the water was decanted, filtered and analyzed. Second, third, fourth and fifth exposures, each 48 hr at room temperature, were included in this procedure. Conductivity and pH were measured at the beginning and end of each exposure period.

Odor Testing

One additional test was included in Phase II, a subjective odor test intended to demonstrate aesthetic acceptability of the processed waste. This type of testing is done routinely at NSF in plastics and wastewater equipment testing programs. A panel of 10-15 persons, screened and trained in odor testing, individually entered a room containing a beaker with 200 ml of water exposed during Phase II to processed wastes. Panelists were asked to rate the room environment as "acceptable" or "unacceptable."

RESULTS

Phase I results are shown in Table III. Data from Phase II are presented as measured levels of constituents leached in Tables IV through VIII for ELT, and in Tables IX and X for the ASTM procedure. It is important to note that measurements at or near analytical detection limits may reflect inconsequential changes in levels of chemicals leached. This should be considered in data interpretation.

Table III. Data from Phase I

Pollutant	Raw Wastes[a]				
	Ni	Cr	Cu	$ZnPO_4$	WTS
Al	30	8.0	50	43	848
Sb	<0.01	0.02	0.1	1.2	2.0
As	0.055	0.02	0.3	<0.1	0.8
Ba	0.3	0.2	0.5	6.5	11.3
Be	<0.001	<0.001	<0.001	<0.01	<0.01
Cd	0.008	0.005	0.01	1.9	0.9
Cr (total)	1.5	1,750	7.3	26	136
Cu	220	7.5	7,500	3.9	39.6
Fe	14.2	0.5	75	3,250	565
Pb	0.4	0.4	4.0	23.4	79
Mn	0.5	0.1	0.7	195	113
Hg	<0.001	<0.001	<0.001	2.6	<0.2
Ni	2,900	40	27	1,040	814
Se	0.013	0.027	0.36	<0.1	0.1
Ag	0.005	0.005	0.063	<0.01	0.09
Tl	<0.003	<0.003	<0.003	<0.03	<0.03
Sn	<0.01	<0.01	0.16	1.6	0.8
V	<0.2	0.2	<0.2	<2.0	<2.0
Zn	15.5	100	17.5	27,625	10,735
TOC	110	45	520	28,000	163,700
COD	397	103	2,000	261,000	313,000
Total Ash	5,900	3,600	39,900	43,200	252,700
NH_3[b]	1.4	1.3	9.3	66	19.2
NO_3 (as N)	1.7	2.2	6.3	1,470	110
P (as P)	0.5	1.4	<0.05	13,520	780
Cl	35.5	17.8	10,650	<5.0	320
CN	9.1	<0.003	5.8	1.6	129
F	3.6	6.5	6.9	3,900	994
SO_4[b]	60	664	482	39	29.4
S^{2-}	<0.003	<0.003	<0.003	0.52	0.28
SO_3	6.0	6.2	18	c	c
SCN[b]	<1.0	<1.0	3.2	13.0	2.3
Cond., μmhos/cm	600	1,450	30,000	490[d]	1,000[d]
pH, units	10.47	6.76	11.82	3.62[d]	7.46

[a]All units in mg/l unless otherwise specified. Pollutants in $ZnPO_4$ and WTS wastes were measured as μg/g but converted to mg/l for this and subsequent tables in this report. Densities were 1.30 and 1.13, respectively. Densities of the plating wastes were essentially 1.0.

[b]Analyses performed on filtrates.

[c]Not measurable; organics interference.

[d]1:1 dilution.

Table IV. Nickel Waste–ELT

Pollutant (mg/l)	Raw Waste[a]	STABLEX Exposure Period (days)			
		1	7	14	28
Al	30	3.0	0.3	<0.005	0.01
Cu	220	<0.002	0.003	<0.002	<0.002
Ni	2,900	<0.005	0.01	<0.005	<0.005
CN	9.1	0.03	0.04	<0.03	<0.03
TOC	110	<1.0	<1.0	1.0	<1.0
COD	397	<25	<25	<25	<25
Odor	N/A[b]	A[c]	A[c]	A[c]	A[c]

[a]From Phase I.
[b]Not applicable (no analysis).
[c]Acceptable.

Table V. Chromium Waste–ELT

Pollutant (mg/l)	Raw Waste[a]	STABLEX Exposure Period (days)			
		1	7	14	28
Cr (total)	1750	0.09	0.1	0.03	0.005
Ni	40	<0.005	0.05	<0.005	<0.005
Zn	100	<0.005	0.6	0.02	<0.005
SO_4	664	47	201	69	d
TOC	45	<1.0	<1.0	<1.0	<1.0
COD	103	<25	<25	<25	<25
Odor	N/A[b]	A[c]	A[c]	A[c]	A[c]

[a]From Phase I.
[b]Not applicable (no analysis).
[c]Acceptable.
[d]Laboratory error.

Table VI. Copper Waste–ELT

Pollutant (mg/l)	Raw Waste[a]	STABLEX Exposure Period (days)			
		1	7	14	28
Al	50	8.0	0.06	<0.005	<0.005
Cu	7,500	0.02	0.02	0.008	<0.002
Se	0.36	0.02	0.005	<0.001	<0.001
Cl	10,650	390	82	7.1	<3.5
CN	5.8	0.04	<0.03	<0.03	<0.03
TOC	520	3.0	1.3	<1.0	<1.0
COD	2,000	<25	28	<25	<25
Odor	N/A[b]	A[c]	A[c]	A[c]	A[c]

[a]From Phase I.
[b]Not applicable (no analysis).
[c]Acceptable.

Table VII. Zinc Phosphate Waste–ELT

Pollutant (mg/l)	Raw Waste[a]	STABLEX Exposure Period (days)			
		1	7	14	28
Cd	1.9	<0.0005	<0.0005	<0.0005	<0.0005
Cr (total)	26	0.3	0.02	<0.002	<0.002
Fe	3,250	0.01	0.03	<0.005	<0.005
Pb	23.4	<0.005	<0.005	<0.005	<0.005
Mn	195	<0.005	0.02	0.03	0.01
Hg	2.6	<0.002	<0.002	0.002	0.002
Ni	1,040	<0.005	0.1	0.07	0.03
Zn	27,625	<0.005	0.9	0.7	0.2
NO_3 (as N)	1,470	0.5	1.1	0.09	<0.09
P (as P)	13,520	0.1	<0.05	<0.05	<0.05
F	3,900	6.3	8.8	8.2	2.6
TOC	28,000	2.9	1.9	<1.0	<1.0
COD	261,000	97	<25	<25	<25
Odor	N/A[b]	A[c]	A[c]	A[c]	A[c]

[a]From Phase I.
[b]Not applicable (no analysis).
[c]Acceptable.

Table VIII. Waste Treatment Plant Sludge–ELT

Pollutant (mg/l)	Raw Waste[a]	STABLEX Exposure Period (days)			
		1	7	14	28
Al	848	2.2	1.7	0.2	0.5
As	0.8	<0.005	<0.005	0.02	0.02
Cr (total)	136	0.07	0.05	0.05	0.002
Fe	565	0.005	<0.005	<0.005	0.01
Pb	79	0.01	<0.005	<0.005	0.006
Mn	113	0.01	0.005	0.006	0.01
Zn	10,735	<0.005	<0.005	0.06	0.2
CN	129	0.03	0.04	<0.03	<0.03
TOC	163,700	78	37	6.1	4.6
COD	313,000	806	209	71	48
Odor	N/A[b]	A[c]	A[c]	A[c]	A[c]

[a]From Phase I.
[b]Not applicable (no analysis).
[c]Acceptable.

Table IX. Raw Zinc Phosphate Waste–ASTM

Pollutant (mg/l)	Raw Waste[a]	Exposure Period (days)				
		2	4	6	8	10
Cd	1.9	0.04	0.008	0.003	0.002	0.007
Cr (total)	26	0.08	0.02	0.02	0.02	0.003
Fe	3,250	10.1	1.1	2.0	0.4	0.2
Pb	23.4	0.05	0.02	0.01	0.01	0.007
Mn	195	4.5	1.6	1.1	0.9	0.8
Hg	2.6	<0.002	<0.002	<0.002	<0.002	<0.002
Ni	1,040	50	14	5.0	5.0	26
Zn	27,625	226	138	112	90	134
NO_3 (as N)	1,470	0.2	0.5	7.7	2.4	0.7
P (as P)	13,520	264	132	104	76	63
F	3,900	188	65	170	65	43
TOC	28,000	4.4	<1.0	<1.0	1.5	<1.0
COD	261,000	47	<25	<25	59	<25
Odor	N/A[b]	A[c]	A[c]	A[c]	A[c]	A[c]

[a]From Phase I.
[b]Not applicable (no analysis).
[c]Acceptable.

Table X. STABLEX Zinc Phosphate Waste–ASTM

Pollutant (mg/l)	Raw Waste[a]	Exposure Period (days)				
		2	4	6	8	10
Cd	1.9	0.001	0.0005	<0.0005	<0.0005	<0.0005
Cr (total)	26	0.06	0.02	0.007	0.003	0.002
Fe	3,250	<0.005	0.03	<0.005	<0.005	0.01
Pb	23.4	0.03	<0.005	<0.005	<0.005	<0.005
Mn	195	0.02	<0.005	0.006	0.009	<0.005
Hg	2.6	0.003	<0.002	<0.002	<0.002	0.002
Ni	1,040	<0.005	0.01	0.01	0.01	0.03
Zn	27,625	<0.005	0.1	0.2	0.3	0.7
NO_3 (as N)	1,470	[c]	6.7	1.9	0.5	<0.09
P (as P)	13,520	<0.05	0.2	0.1	0.3	0.2
F	3,900	50	30	8.1	5.9	2.9
TOC	28,000	8.2	1.8	<1.0	1.4	<1.0
COD	261,000	67	41	<25	<25	<25
Odor	N/A[b]	A[d]	A[d]	A[d]	A[d]	A[d]

[a]From Phase I.
[b]Not applicable (no analysis).
[c]Laboratory error.
[d]Acceptable.

Tables XI and XII present the data from Tables IX and X, reported in accordance with the procedure specified by ASTM, where leaching values (L) are defined as milligrams of pollutants leached into water per gram of sample exposed.

Conductivity and pH data during ELT for processed zinc phosphate waste are shown in Figures 1 and 2, respectively. Similar plots for the other four wastes appear in Appendix B. Conductivity and pH measurements from the ASTM procedure (raw and stabilized zinc phosphate waste) are shown in Tables XIII and XIV, respectively. It is important to note that no chemical buffering was introduced to the test system; therefore, the low-pH environment present during initial exposure quickly shifted (less than 24 hr) to alkaline conditions as a result of exposure to the processed waste. It is assumed this shift could be expected when processed waste is exposed to rainfall or runoff in land disposal/reclamation applications.

DISCUSSION

Equilibrium Leaching Test

In the ELT, it is expected that any constituent available in an exposed sample will leach to the extractant medium (water) until an equilibrium condition is reached. The reaction kinetics are greater with fresh extractant water because levels of constituents of interest are essentially zero. This effect was monitored in the study by daily conductivity measurements. Figure 3, for example, which contains conductivity data for the STABLEX processed chromium waste, shows that changing the water at days 7 and 14 during the test induced higher levels of leaching (i.e., ionic strength increased), but an equilibrium condition was quickly reached. Considering that in this test, the sample is pulverized, providing up to 5000 times the surface area of exposure which should be expected with land disposal, and exposure was accomplished with changes of water at relatively low pH (4.5). ELT is easily established as an accelerated test method.

Regulated Pollutants

Data relating to leaching of the nine inorganic chemicals for which maximum contaminant levels (MCL) are listed in the EPA National Interim Primary Drinking Water Regulations are impressive. These chemicals and their associated MCL are identified in Table XV.

On December 18, 1978, EPA published "Proposed Guidelines and Regulations and Proposal on Identification and Listing of Hazardous Wastes" [4]. A solid waste is defined as "hazardous" if, according to the two methods specified in the Proposed Guidelines, the extract from exposure of a representative sample of the waste contains levels of any contaminants in excess of ten times the EPA National Interim Primary Drinking Water Regulations. (It is noted in the Proposed Regulations that MCL for hazardous wastes will be changed

Table XI. ASTM Results Expressed as Leaching Values–Raw Zinc Phosphate Waste

Pollutant "L"[a]	ASTM Test Exposure Period (days)				
	2	4	6	8	10
Cd	5.5×10^{-4}	1.1×10^{-4}	4.1×10^{-5}	2.8×10^{-5}	9.7×10^{-5}
Cr (total)	1.1×10^{-3}	2.8×10^{-4}	2.8×10^{-4}	2.8×10^{-4}	4.0×10^{-5}
Fe	1.4×10^{-1}	1.5×10^{-2}	2.8×10^{-2}	5.5×10^{-3}	2.8×10^{-3}
Pb	7.0×10^{-4}	2.8×10^{-4}	1.4×10^{-4}	1.4×10^{-4}	1.0×10^{-4}
Mn	6.2×10^{-2}	2.2×10^{-2}	1.5×10^{-2}	1.2×10^{-2}	1.1×10^{-2}
Hg	$<3.0 \times 10^{-5}$	$<3.0 \times 10^{-5}$	$<3.0 \times 10^{-5}$	$<3.0 \times 10^{-5}$	$<3.0 \times 10^{-5}$
Ni	6.9×10^{-1}	1.9×10^{-1}	6.9×10^{-2}	6.9×10^{-2}	3.6×10^{-1}
Zn	3.1×10^{0}	1.9×10^{0}	1.5×10^{0}	1.2×10^{0}	1.9×10^{0}
NO_3 (as N)	2.7×10^{-3}	6.8×10^{-3}	1.1×10^{-1}	3.3×10^{-2}	9.7×10^{-3}
P (as P)	3.6×10^{0}	1.8×10^{0}	1.4×10^{0}	1.1×10^{0}	8.7×10^{-1}
F	2.6×10^{0}	9.0×10^{-1}	2.4×10^{0}	9.0×10^{-1}	5.9×10^{-1}
TOC	6.1×10^{-2}	$<1.4 \times 10^{-2}$	$<1.4 \times 10^{-2}$	2.1×10^{-2}	$<1.4 \times 10^{-2}$
COD	6.5×10^{-1}	$<3.4 \times 10^{-1}$	$<3.4 \times 10^{-1}$	8.1×10^{-1}	$<3.4 \times 10^{-1}$

[a] $L = \frac{C}{250\,S}$ where L = leaching value
C = concentration of parameter of interest, mg/l
S = solid content in sample = 0.29 g/g (raw waste)

Table XII. ASTM Results Expressed as Leaching Values–STABLEX Zinc Phosphate Waste

Pollutant "L"[a]	ASTM Test Exposure Period (days)				
	2	4	6	8	10
Cd	4.0×10^{-6}	2.0×10^{-6}	$<2.0 \times 10^{-6}$	$<2.0 \times 10^{-6}$	$<2.0 \times 10^{-6}$
Cr (total)	2.4×10^{-4}	8.0×10^{-5}	2.8×10^{-5}	1.2×10^{-5}	8.0×10^{-6}
Fe	$<2.0 \times 10^{-5}$	1.2×10^{-4}	$<2.0 \times 10^{-5}$	$<2.0 \times 10^{-5}$	4.0×10^{-5}
Pb	1.2×10^{-4}	$<2.0 \times 10^{-5}$	$<2.0 \times 10^{-5}$	$<2.0 \times 10^{-5}$	$<2.0 \times 10^{-5}$
Mn	8.0×10^{-5}	$<2.0 \times 10^{-5}$	2.4×10^{-5}	3.6×10^{-5}	$<2.0 \times 10^{-5}$
Hg	1.2×10^{-5}	$<8.0 \times 10^{-6}$	$<8.0 \times 10^{-6}$	$<8.0 \times 10^{-6}$	8.0×10^{-6}
Ni	$<2.0 \times 10^{-5}$	4.0×10^{-5}	4.0×10^{-5}	4.0×10^{-5}	1.2×10^{-4}
Zn	$<2.0 \times 10^{-5}$	4.0×10^{-4}	8.0×10^{-4}	1.2×10^{-3}	2.8×10^{-3}
NO_3 (as N)	[b]	2.7×10^{-2}	7.7×10^{-3}	2.0×10^{-3}	$<3.6 \times 10^{-4}$
P (as P)	$<2.0 \times 10^{-4}$	8.0×10^{-4}	4.0×10^{-4}	1.2×10^{-3}	8.0×10^{-4}
F	2.0×10^{-1}	1.2×10^{-1}	3.2×10^{-2}	2.4×10^{-2}	1.2×10^{-2}
TOC	3.3×10^{-2}	7.2×10^{-3}	$<4.0 \times 10^{-3}$	5.6×10^{-3}	$<4.0 \times 10^{-3}$
COD	2.7×10^{-1}	1.6×10^{-1}	$<1.0 \times 10^{-1}$	$<1.0 \times 10^{-1}$	$<1.0 \times 10^{-1}$

[a] Term S in expression for "L" (see Table XI) assumed = 0.95 g/g (STABLEX).
[b] Laboratory error.

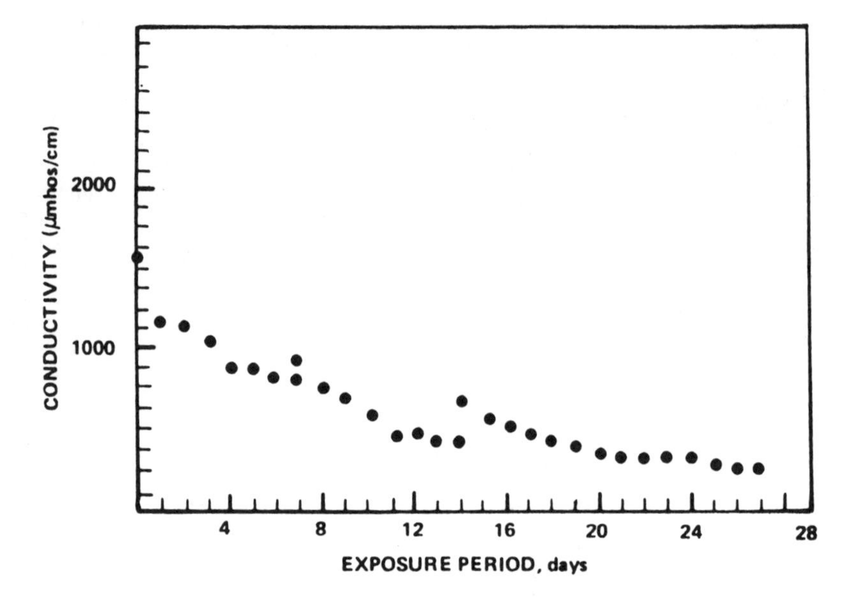

Figure 1. Conductivity during ELT of $ZnPO_4$ processed waste.

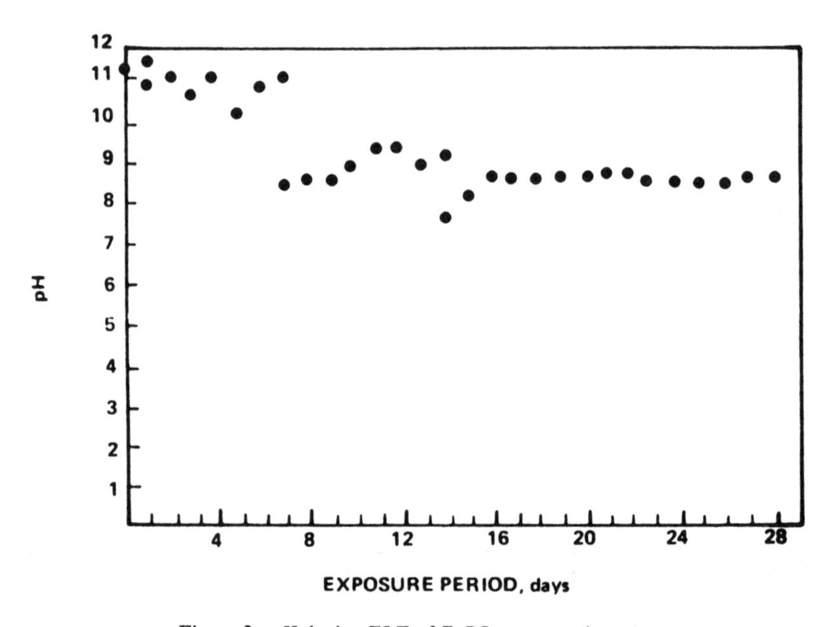

Figure 2. pH during ELT of $ZnPO_4$ processed waste.

Table XIII. Conductivity in ASTM Test

Sample (μmhos/cm)	Exposure Period (days)									
	2		4		6		8		10	
	I	F	I	F	I	F	I	F	I	F
Control	44.5	9.5	30.0	3.4	27.8	3.7	21.9	3.7	21.3	4.9
Raw $ZnPO_4$ Waste	44.5	2,300	385	820	233	535	185	432	105	357
STABLEX	44.5	1,970	470	830	305	475	229	368	203	318

Table XIV. pH in ASTM Test

Sample (pH units)	Exposure Period (days)									
	2		4		6		8		10	
	I	F	I	F	I	F	I	F	I	F
Control	3.6	4.6	4.1	5.4	4.2	4.3	4.4	5.5	4.0	6.4
Raw $ZnPO_4$ Waste	3.6	4.0	4.6	4.2	4.7	4.5	4.9	4.4	4.5	4.7
STABLEX	3.6	9.6	5.9	8.7	5.8	8.0	5.9	7.6	5.5	7.9

Table XV. Maximum Contaminant Levels for Inorganic Chemicals in National Interim Primary Drinking Water Regulations

Contaminant	Level (mg/l)
As	0.05
Ba	1.
Cd	0.010
Cr (total)	0.05
Pb	0.05
Hg	0.002
NO_3 (as N)	10.
Se	0.01
Ag	0.05

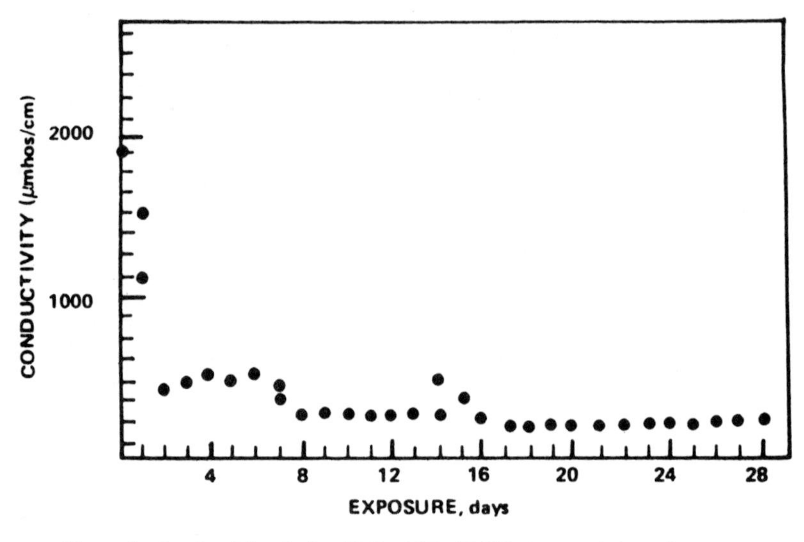

Figure 3. Conductivity during ELT of STABLEX processed chromium waste.

to reflect changes in the Drinking Water Regulations, and that Water Quality Criteria under the Clean Water Act may also be used in setting extractant levels.)

Although neither extraction procedure used in this study is cited in the Proposed Regulations, one or both are at least as severe as either of the proposed methods. All nine inorganic chemicals in the Proposed Regulations were included in Phase I, i.e., measured in the raw wastes (Table III). Seven of the nine (As, Cd, Cr, Pb, Hg, Se and NO_3) were included in Phase II, i.e., measured in extractant water exposed to one or more of the processed wastes. Barium and silver were not selected for Phase II because of their relatively low levels in the raw wastes.

Arsenic

Arsenic was present at relatively low levels in each of the raw wastes (0.02 mg/l in Cr waste to 0.8 mg/l in WTS). It was measured in Phase II only in leachate from the waste treatment plant sludge. As shown in Figure 4, the level after each period of exposure was less than the MCL established for drinking water.

Cadmium

The highest level of Cd in any of the raw wastes was 1.9 mg/l in the paint priming waste ($ZnPO_4$). Leachate measurements from the STABLEX exposures were reported as <0.0005 mg/l for each period of exposure included in the study (Figure 5).

Chromium

Total chromium was measured in Phase II on three processed wastes: Cr, $ZnPO_4$ and WTS. As shown in Figure 6, total Cr in the raw wastes varied from 26 mg/l ($ZnPO_4$ waste) to 1750 mg/l (Cr waste). At the end of a single 24-hr exposure, all samples in Phase II contained less than the proposed MCL for hazardous wastes; after the third exposure (end of 14-day period), no level exceeded the MCL for drinking water.

Lead

Lead was measured at 23.4 and 79 mg/l in the raw $ZnPO_4$ and WTS wastes, respectively. After 24 hr (and all subsequent periods) of exposure of the processed wastes, levels were reduced to less than the MCL for drinking water (Figure 7).

Mercury

Mercury, present in the $ZnPO_4$ waste at 2.6 mg/l was similarly reduced to equal to or less than the drinking water MCL for all exposure periods tested (Figure 8).

Selenium

Selenium was present in the raw copper waste at 0.36 mg/l, reduced to less than the MCL for hazardous wastes in the first 24-hr exposure, and to less than the drinking water MCL for subsequent exposures (Figure 9).

Nitrate

Nitrate was reduced from 1470 mg/l in the raw $ZnPO_4$ waste to less than the drinking water MCL in all exposures of the processed waste, as shown in Figure 10.

Amphoteric Hydroxides

In addition to inorganic chemicals with drinking water regulatory significance, there was concern for the leaching potential of metals with amphoteric hydroxides, e.g., Al and Zn. Al was present in the raw Ni, Cu and WTS wastes at levels of 30, 50 and 848 mg/l, respectively. Zinc was present in raw Cr, $ZnPO_4$ and WTS wastes at respective levels of 100, 27,625 and 10,735 mg/l. The data show that amphoteric characteristics alone do not contribute significantly to the leaching tendencies of STABLEX-processed wastes.

"Excessive" Levels

It is interesting to note that all of the cations present in one or more of the raw wastes at very high levels, i.e., Ni, Cu, Fe and Mn, were effectively bound by the polymer. Except for CN and P, the anions were less effectively contained by STABLEX. Most notable among the relatively high levels of

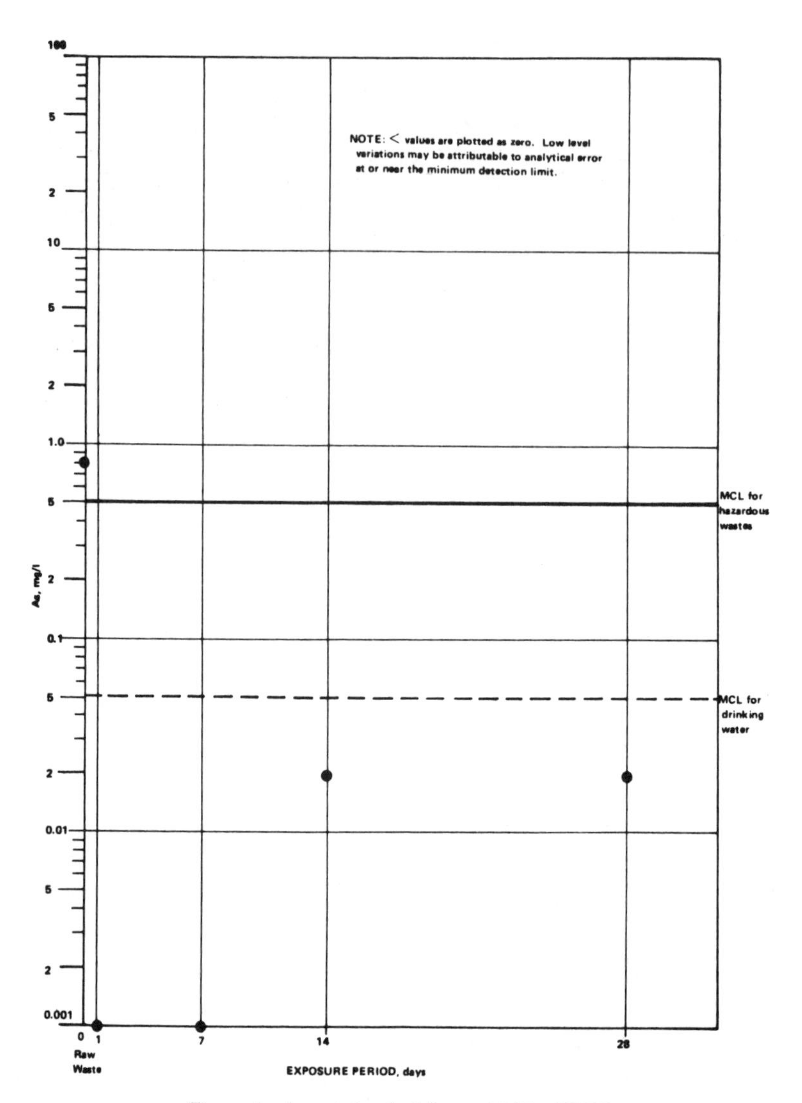

Figure 4. Arsenic leached from stabilized WTS.

anions leached were SO_4, Cl and F, in that order. Although the concentrations of leached Cl and F were significantly higher than other inorganic chemicals monitored in Phase II, less than 0.5% of the levels measured in raw waste was leached (total) from the stabilized wastes. Nearly 5% of the SO_4 present in the raw Cr waste was leached from the STABLEX.

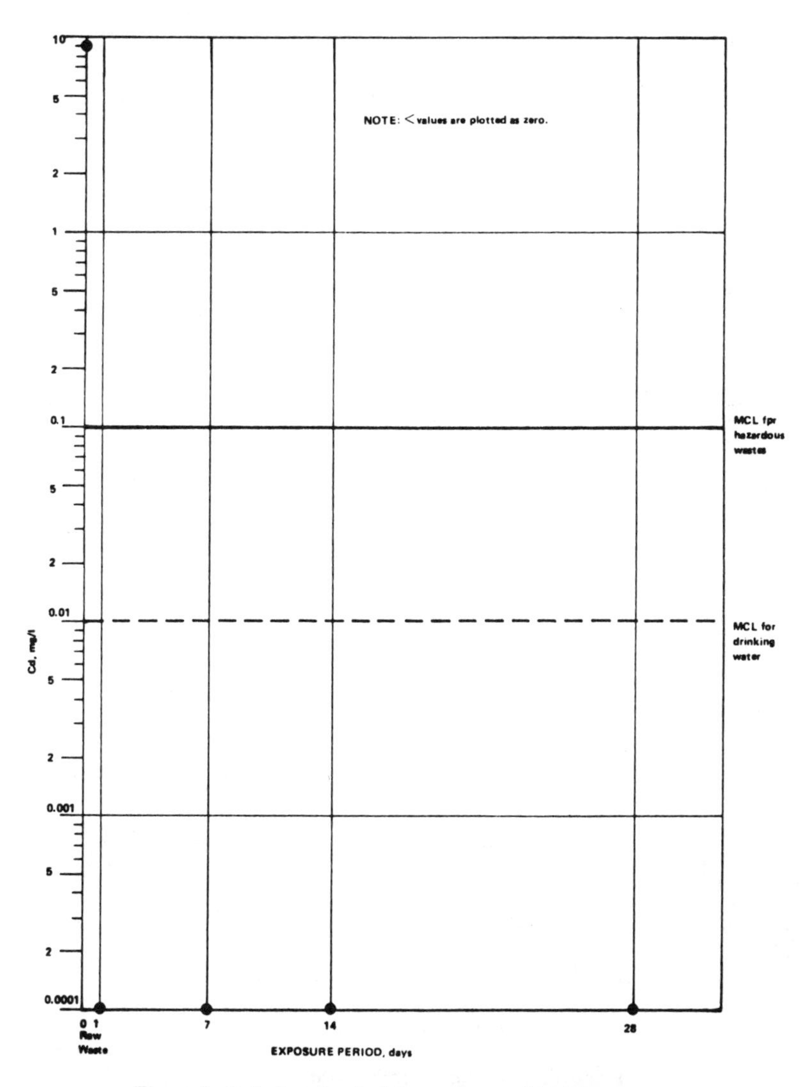

Figure 5. Cadmium leached from processed $ZnPO_4$ waste.

ASTM Test

It is readily apparent from Tables IX and X that much higher levels of pollutants would be leached to the environment from raw than from STABLEX-processed wastes.

Data from ELT and the ASTM tests compare favorably, despite significant differences in the two procedures. Samples for ELT are pulverized; samples

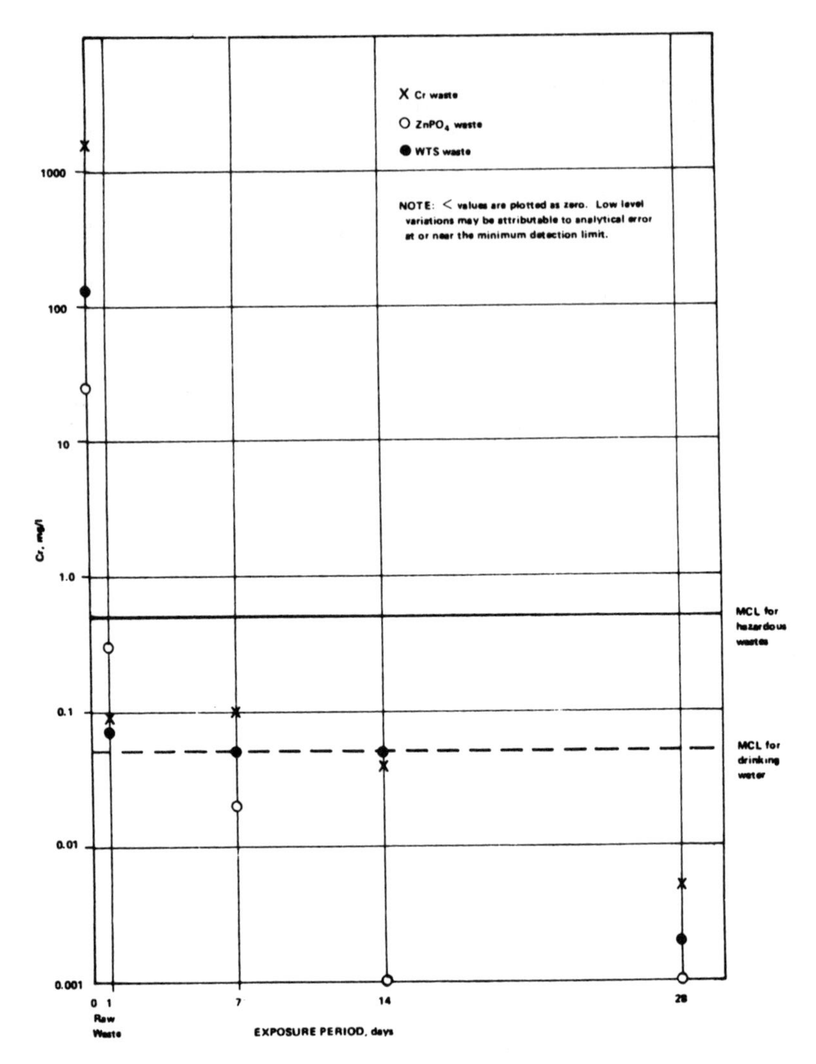

Figure 6. Total chromium leached during ELT.

for the ASTM method are used as solid core samples initially, although they may fracture during testing. Dilutions also vary. One part sample to ten parts water by weight is used in ELT; one part sample to four parts water in ASTM. Mixing time and methods and exposure periods also differ between the two methods. Data in Tables VII and X are leachate measurements for STABLEX-processed $ZnPO_4$ waste using ELT and ASTM, respectively. (Although the data in Table X appear highly variable, i.e., concentration is not necessarily reduced with consecutive exposures, round-robin testing, required by ASTM

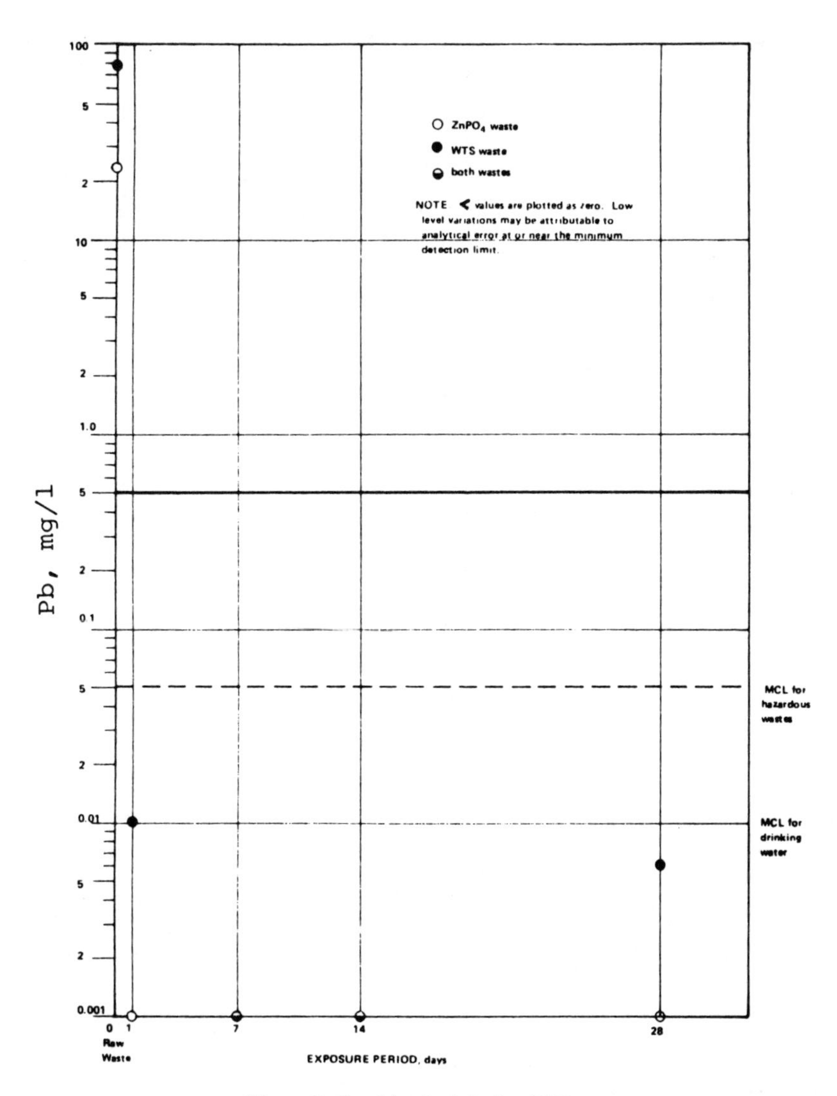

Figure 7. Lead leached during ELT.

before a Proposed Method is adopted as a standard, is reported to produce considerable variability from laboratory to laboratory.)

Odor Testing

None of the leachate samples tested was rated "unacceptable" with respect to odor by majority panelist opinion. This is particularly noteworthy because the odor from the raw waste treatment plant sludge was nearly overpowering.

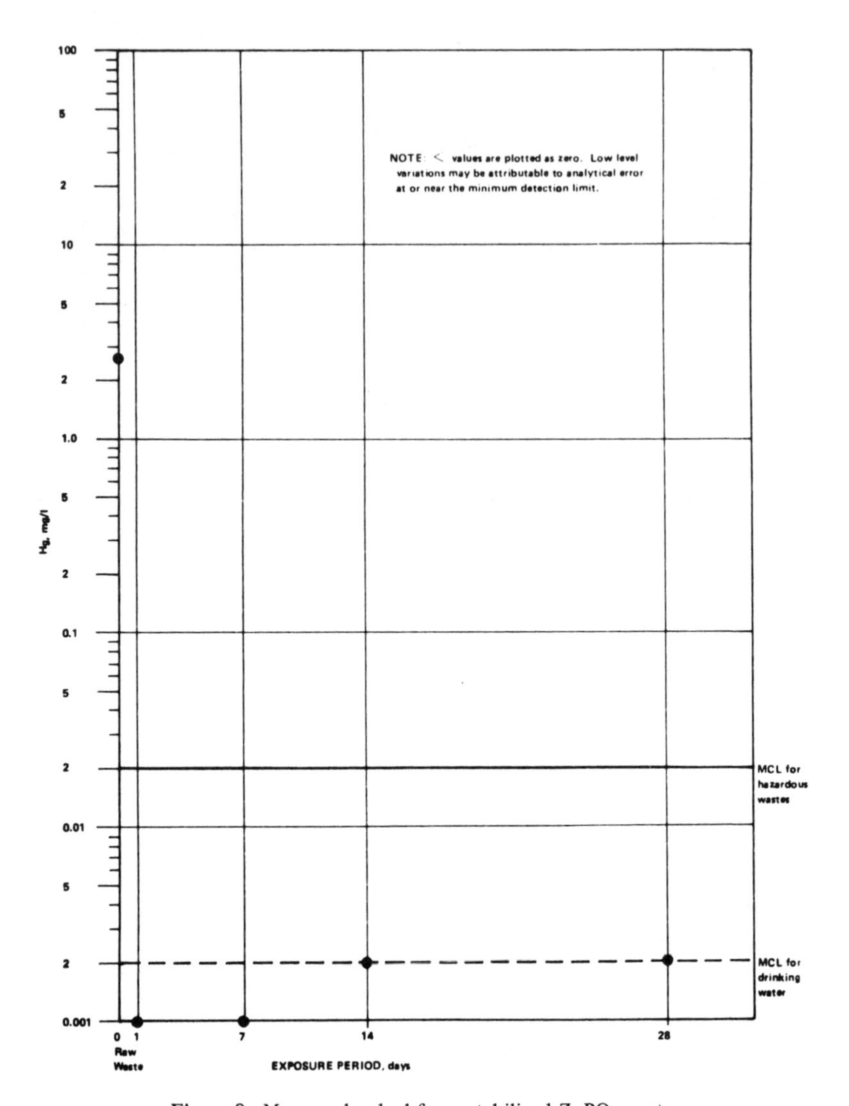

Figure 8. Mercury leached from stabilized $ZnPO_4$ waste.

CONCLUSIONS

Five samples of typical automotive wastes which contained a wide qualitative and quantitative range of inorganic pollutants were shown to be effectively stabilized when converted to STABLEX.

Inorganic chemicals listed in the EPA National Interim Primary Drinking

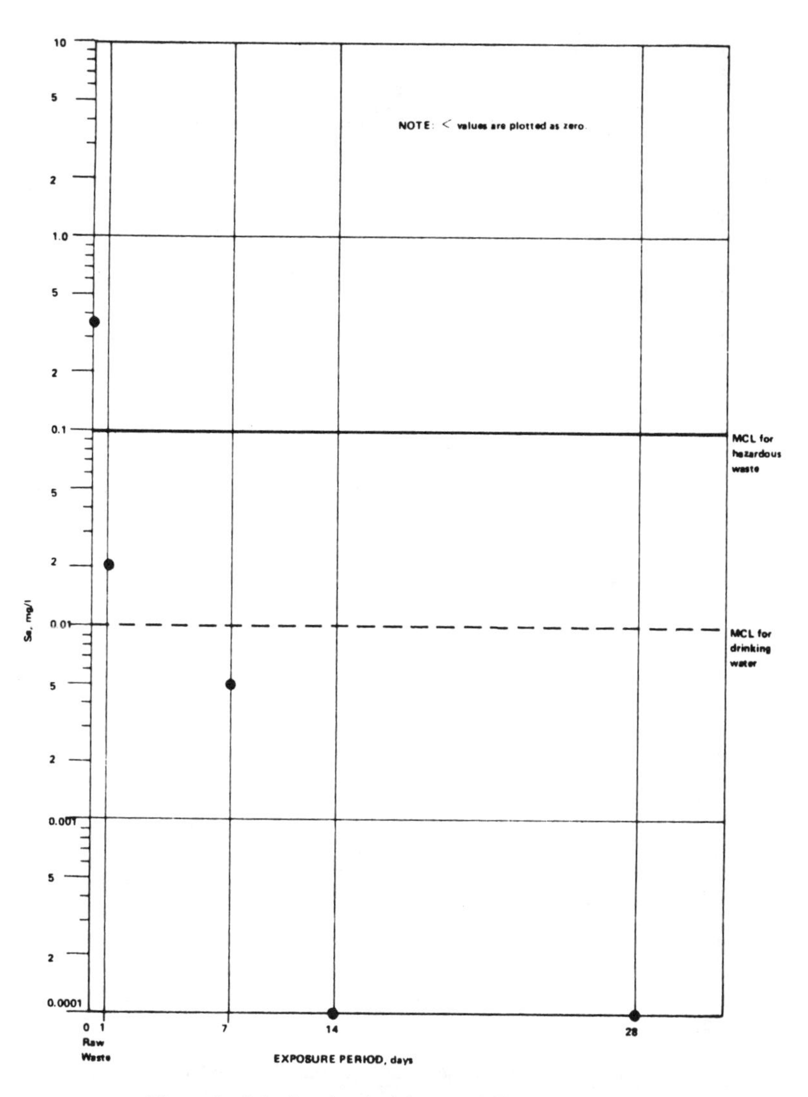

Figure 9. Selenium leached from stabilized Cu waste.

Water Regulations and Proposed Regulations for Hazardous Wastes were present in the raw wastes at varying and, in some cases, very high levels. Leachate testing of the STABLEX-processed wastes produced no level of any of the regulated chemicals greater than the MCL proposed for hazardous waste disposal. Five of the chemicals–As, Cd, Hg, Pb and NO_3–were less than MCL in the Drinking Water Regulations in all samples tested.

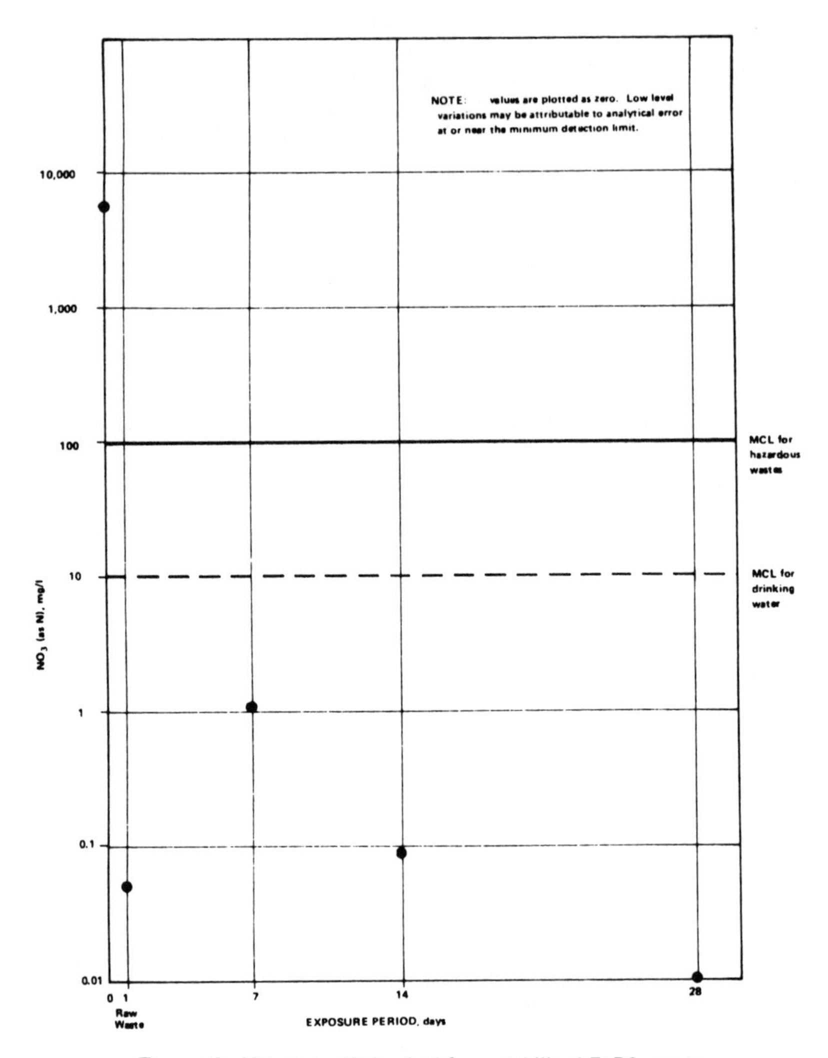

Figure 10. Nitrate (as N) leached from stabilized $ZnPO_4$ waste.

The presence of relatively high levels of organic constituents did not impede the effectiveness of STABLEX in binding the inorganic pollutants.

With adequate site-selection and process operation, there is no indication from this study that STABLEX processed automotive wastes would leach significant levels of hazardous inorganic pollutants to the environment when exposed to naturally occurring groundwater (pH $\geqslant$4.5).

ACKNOWLEDGMENTS

The services of those who participated in planning and conducting this study are recognized with sincere appreciation: Dennis Tierney and Diane M. Carlson, Michigan Department of Natural Resources; Dr. Robert B. Pojasek, Energy Resources Company, Inc.; John T. Schofield and Steven I. Taub, Stablex Corporation; and Peter J. Derrett and William P. Dolan, Stablex-Ruetter, Inc.

REFERENCES

1. U.S. Environmental Protection Agency. "National Interim Primary Drinking Water Regulations," EPA-570/9-76-003, Washington, DC (1977).
2. Taub, S. I., and B. K. Roberts. "Leach Testing of Chemically Stabilized Waste," paper presented at the Third Annual Conference on Treatment and Disposal of Industrial Wastewaters and Residues, Houston, Texas, 1978.
3. U.S. Environmental Protection Agency. "Hazardous Waste Proposed Guidelines and Regulations and Proposal on Identification and Listing," 43 CFR 234, Part IV, December 18, 1978.
4. "Proposed Method D19:12, Method A," *Annual Book of ASTM Standards* (Philadelphia, PA: American Society for Testing and Materials, 1979).

APPENDIX A

ANALYTICAL METHODS AND SPECIFICATIONS

1. Metals
   - Method: Atomic absorption spectrophotometry (AA)
   - Ref: *Standard Methods*, 14th ed., p. 147
   - Matrix: Leachates preserved with 0.2% nitric acid
   - Instruments: Perkin-Elmer (P.E.) AA, Model 560
     - P.E. heated graphite atomizer (HGA), Model 2200
     - P.E. autosampler, Model AS1
     - Argon gas
   - Settings: AA
     - Wavelength, primary for each metal;
     - Slit, as specified by P.E.
     - Furnace (HGA)
       - See Table A-I
   - Sample Size: 20 μl
2. Total Organic Carbon
   - Method: Automated high-temperature combustion
   - Instruments: Hewlett Packed CHN Analyzer (raw wastes)
     - Beckman 915TOC Analyzer (STABLEX)
   - Minimum Detection Limit (MDL): 1 mg/l C

Table A-I. HGA Specifications

Metal	Dry Temp. (°C)	Dry Time (sec)	Char Temp. (°C)	Char Time (sec)	Atomization Temp. (°C)	Atomization Time (sec)	MDL (mg/l)
Al	110	20	1410	32	2700	5	0.005
Sb	110	27	825	22	2700	6	0.010
As	110	20	290	31	2700	7.5	0.005
Ba	110	23	1100	36	2800	9	0.2
Be	110	20	1200	30	2800	9	0.001
Cd	110	20	280	32	2100	6	0.0005
Cr	120	22	1100	30	2700	9	0.002
Cu	120	22	900	30	2700	9	0.002
Fe	120	22	1100	30	2700	10	0.005
Pb	120	22	700	30	2300	9	0.005
Mn	120	22	1100	30	2710	9	0.005
Ni	120	23	1000	30	2700	9	0.005
Se	110	22	375	21	2700	9	0.001
Ag	110	20	400	30	2700	9	0.005
Tl	110	20	400	30	2300	10	0.003
Sn	110	20	700	32	2700	6	0.010
V	110	20	1700	30	2800	10	0.20
Zn	130	22	500	30	2500	6	0.005
Hg[a]	–	–	–	–	–	–	0.002

[a]Cold vapor method, Perkin-Elmer Mercury Analysis System 303-0830.

3. Chemical Oxygen Demand
   - Method: Dichromate reflux
   - Ref: *Standard Methods*, 14th ed., p. 550
   - MDL: 25 mg/l
4. Total Ash
   - Method: Gravimetric with ignition of residue
   - Ref: *Standard Methods*, 14th ed., p. 96
5. Chloride
   - Method: Ion-selective electrode
   - Sensor: Orion chloride electrode No. 9477.
   - MDL: 3.5 mg/l
6. Cyanide
   - Method: Potentiometric titration
   - Sensor: Orion silver-sulfide electrode, Model 94-16 with distillation step for raw wastes
   - MDL: 0.03 mg/l

7. Fluoride
   - Method: Ion-selective electrode
   - Ref: *Standard Methods*, 14th ed., p. 389
   - Sensor: Orion fluoride electrode Model 94-09A with distillation step for raw wastes
   - MDL: 0.2 mg/l
8. Sulfate
   - Method: Gravimetric with ignition of residue
   - Ref: *Standard Methods*, 14th ed., p. 493
   - MDL: 10 mg/l
9. Sulfide
   - Method: Ion-selective electrode
   - Sensor: Orion silver-sulfide electrode, Model 94-16
   - MDL: 0.003 mg/l
10. Sulfite
    - Method: Titration
    - Ref: *Standard Methods*, 14th ed., p. 508
    - MDL: 2.0 mg/l SO_3
11. Thiocyanate
    - Method: Colorimetry
    - Ref: *Standard Methods*, 14th ed., p. 383
    - MDL: 1.0 mg/l
12. Ammonia
    - Method: Direct Nesslerization
    - Ref: *Standard Methods*, 14th ed., p. 412
    - MDL: 0.05 mg/l
13. Nitrate
    - Method: Brucine
    - Ref: *Standard Methods*, 14th ed., p. 427
    - MDL: 0.1 mg/l N
14. Phosphorus
    - Method: Ascorbic acid with persulfate digestion
    - Ref: *Standard Methods*, 14th ed., pp. 476, 479
    - MDL: 0.05 mg/l
15. Conductivity
    - Method: Electrical conductivity
    - Ref: *Standard Methods*, 14th ed., p. 71
    - Instruments: Conductivity Bridge, Yellow Springs Instrument Company (YSI), Model 31
    - MDL: 1 μmho/cm
16. pH
    - Instruments: Corning digital pH meter, Model 110, with glass and calomel electrodes
    - Ref: *Standard Methods*, 14th ed., p. 461
    - Buffers: P.E. Certified Buffer Tablets, pH 4.00 and 7.02

APPENDIX B

CONDUCTIVITY AND pH FOR Ni, Cr, Cu AND WTS

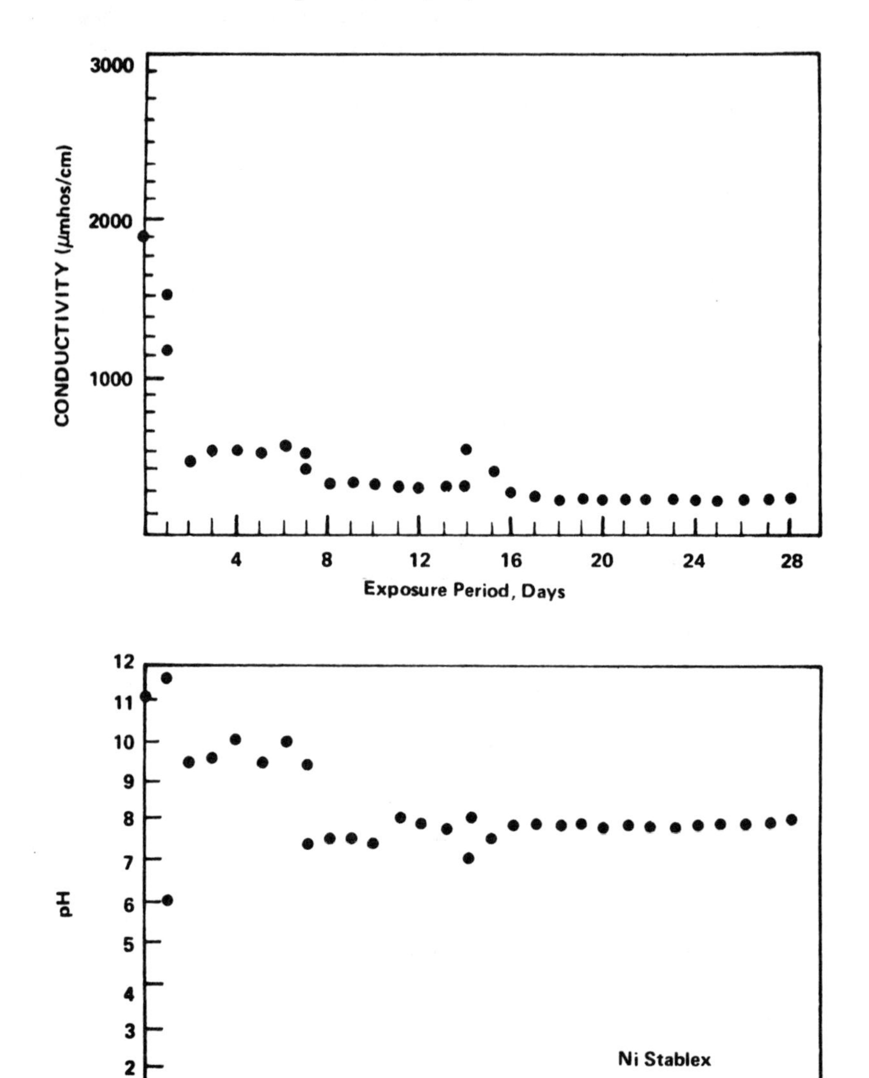

Figure B-1. Ni STABLEX.

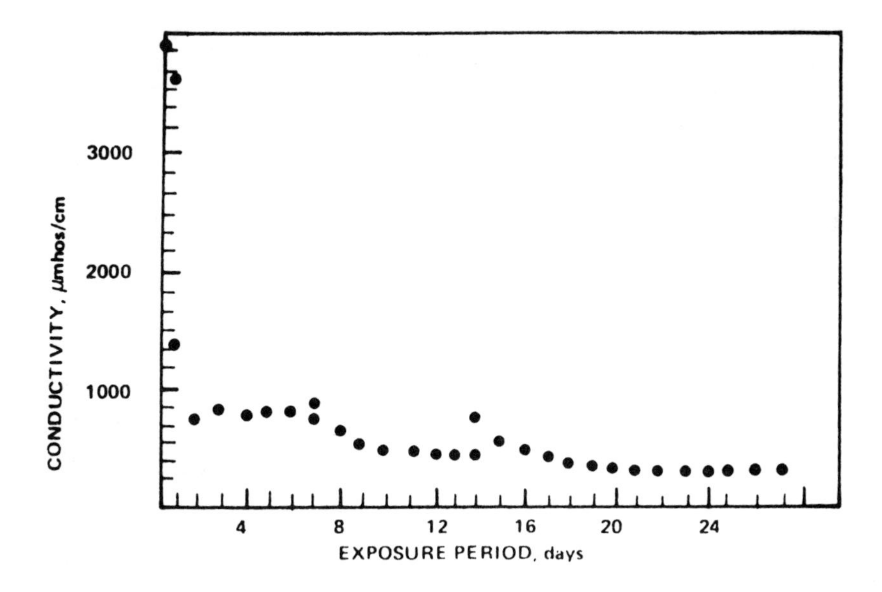

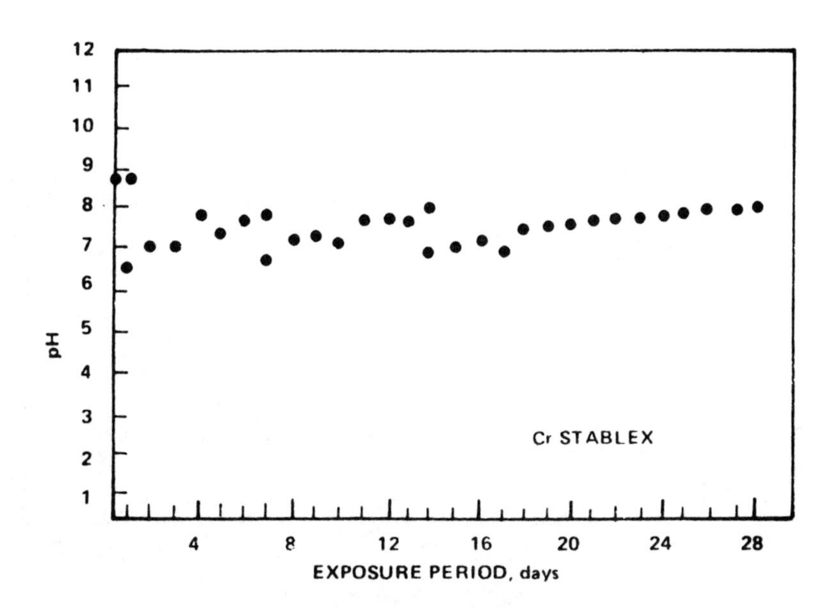

Figure B-2. Cr STABLEX.

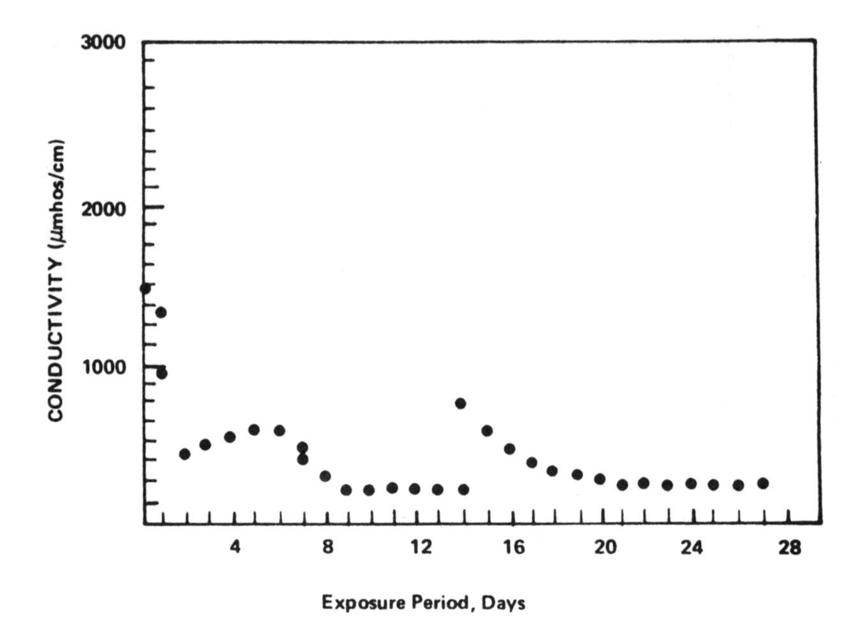

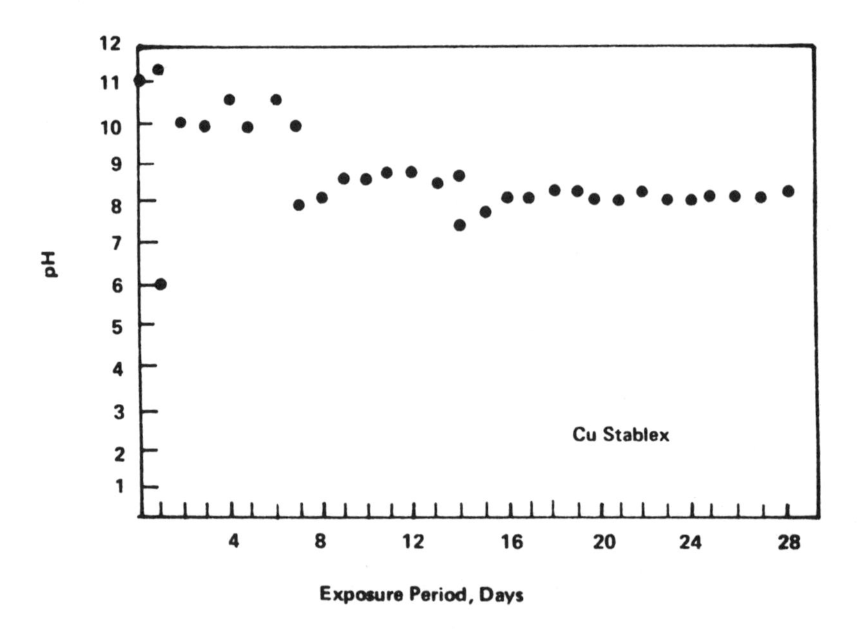

Figure B-3. Cu STABLEX.

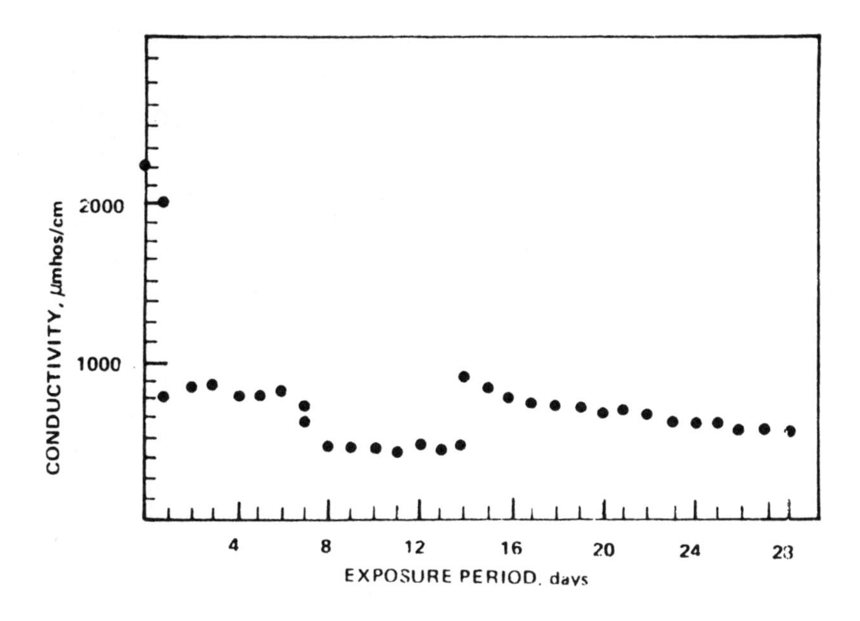

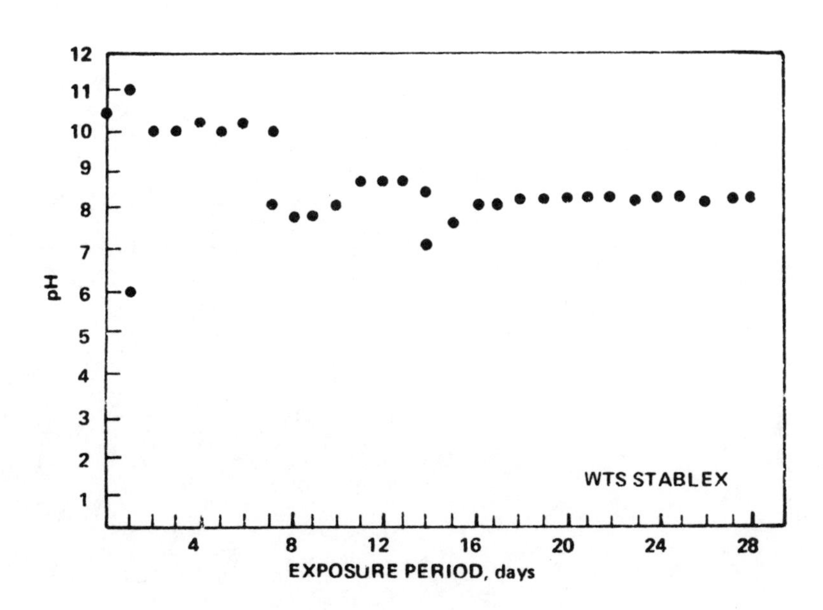

Figure B-4. WTS STABLEX.

CHAPTER 8

LAND RECLAMATION WITH STABILIZED INDUSTRIAL WASTE

John T. Schofield
Stablex Corporation
Radnor, Pennsylvania

In January 1979 Stablex Corporation* of Radnor, PA, announced plans to establish in the State of Michigan the world's largest land reclamation project using stabilized industrial waste. The project, called "Groveland Waste Management and Land Reclamation Center," involved the erection of a facility costing $7-10 million to receive and process 500,000 tons of industrial waste each year to produce STABLEX(TM), an environmentally safe synthetic rock which will be used to reclaim a 200-ac area ravaged by sand and gravel extraction.

The siting of a facility to convert industrial waste to a useful end product has been welcomed by industry in an area characterized by high industrial activity but lacking in adequate waste disposal facilities.

The location of waste disposal facilities has, however, increasingly become a matter of great public concern. A number of major incidents, typified by the Love Canal disaster in New York State, have justifiably turned public opinion against the dumping of untreated hazardous wastes in landfill sites.

Additionally, the U.S. Environmental Protection Agency (EPA), in its Proposed Guidelines covering hazardous waste disposal published in December 1978 stated:

> The agency prefers chemical, physical and biological treatment rather than disposal techniques such as landfills as a means of waste management because such treatment can detoxify waste and thus reduce the potential for human health or environmental damage. Treatment also reduces the burden that present disposal practices place on future resources [1].

---

*The Sealosafe service includes a process protected by patents and patent applications in the United States and overseas and SEALOSAFE and STABLEX are trademarks of the Stablex Group of Companies.

The Groveland plan to treat wastes to produce a useful end product is therefore receiving major attention from federal regulatory authorities, state agencies, environmental groups and members of the public as representing a unique breakthrough in disposal technology.

The SEALOSAFE(SM) service, although newly introduced into North America, is a proven, patented technology which has been in use for several years in the U.K. and more recently in Japan. It is interesting to trace the development of this technology, and in particular its involvement with land reclamation.

In countries like the Netherlands, typical of a small land mass which is highly populated, land reclamation has been practiced for hundreds of years. Because of its low-lying nature, the Netherlands has always been unable to dispose of material to river or sea, but sterilization of land by dumping untreated industrial waste was also banned as being counter to the policy of land reclamation of that country. Fortunately, the Netherlands is attached to the larger land mass of the continent in Europe and disposal of these wastes by trucking them into Belgium or West Germany has provided them with an outlet.

In the case of England and Japan, no such disposal outlet was available. Highly industrialized islands such as England and Japan have in the past, probably more than any other country, disposed of the largest proportion of their waste to rivers and the sea. Introduction of legislation to close these outlets has meant the increased deposition of untreated industrial wastes on to land. The small, highly populated island land masses were unable to meet the demands being placed on them; therefore, legislation was introduced at an early stage to control deposition to land. In England, the first work associated with investigating the problem commenced in 1964 and was followed only a few years later by Japan. This contrasts markedly with countries having large land masses, such as the U.S., where real concern about disposal to land has only arisen in the late 1970s.

Research into the problem of detoxifying industrial wastes therefore commenced in England in the late 1900s, and it was at this time that SEALOSAFE was invented. The technology involves converting hazardous and toxic waste into an inert, environmentally safe synthetic rock, STABLEX, the properties of which recommend the product as being suitable as a land reclamation material. The ability, therefore, to turn otherwise unusable materials into a product which was in great demand in England, resulted in early commercial exploitation of this technology.

The first commercial facility in the U.K. was brought on stream in 1974 and today has an operational capability of dealing with 80,000 ton/year of industrial waste. To establish the facility, all the requirements of legislation introduced into the U.K. in 1972 and 1974 had to be met. This legislation had a fundamental effect on the way in which the technology was deployed commercially. The location of this facility is just north of Birmingham, England, which is an area of high industrial activity. The choice of site, however, was governed by locating within a derelict area which local residents wished to see reclaimed.

Ideal land reclamation materials would be safe, inert, solid, easy to place, nonbiodegradable, nonflammable, no odor, no taste, not attractive to birds and rodents and capable of supporting structures or topsoil for return to agricultural use. The material should also be available in sufficiently large quantities that land reclamation projects can be undertaken and completed within a time scale of visible results. To be able to suggest to local communities that such a product existed free of charge seemed impossible. To suggest further that the location of the facility to produce this material would provide jobs to the local community and, in addition, would provide a substantial tax contribution, provided the motivation for community acceptance of this first commercial facility which is now in its sixth year of successful operation.

The demand on the facility from industrial waste producers was so high that within two years a second facility with a design capacity of 120,000 ton/yr was in development.

Before permission was granted for this second facility, a thorough investigation of the process and product was carried out by the county authority where the plant would be located. Only when the county authority was satisfied that the process and product were environmentally safe, having subjected the product to substantial testing in the public health laboratories and taken due note of the successful operation to date of the first commercial facility, was permission granted and the facility commissioned.

In 1977 application was made to build a third facility in England with the capability of handling 400,000 ton/yr of industrial waste. Before permission was granted to construct this facility, the U.K. government appointed one of its laboratories to carry out a thorough investigation into the process and product. Detailed technical investigation by the investigating officer coupled with a demonstration of the technical and management competence of the company to safely handle a wide range of wastes since 1974 provided the necessary evidence of the unique nature and safety aspects of SEALOSAFE technology. This third facility, located in an extremely derelict area east of London, England, was welcomed by local residents as providing the only known method of restoring to community use many acres of land laid derelict by extraction of chalk for cement manufacture, without any burden on the local taxpayer.

Applications to build a fourth facility with a capability of 200,000 ton/yr are now in progress and construction is expected to commence in the not-too-distant future.

The four facilities, built as regional treatment centers will effectively serve the whole of England.

The STABLEX produced at these English facilities looks like ready-mixed concrete which is easily pumpable. The methods used for land reclamation involve pumping the product to the reclamation site, which is usually irregular due to selective extraction methods employed. The product, therefore, is able to flow over the many irregularities, and reclamation can take place on an orderly basis. The reclamation area is divided into cells and as cells are restored they are returned to other uses.

The most recent facility to be brought on line is located in Kariya City, near Nagoya in Japan. This development, the first commercial plant in Japan to convert hazardous waste into an environmentally safe product, was permitted following appropriate independent testing of the product in Japan. The facility, operated by San Eigumi, a subsidiary of Toyota-Shoki, commenced operation in early 1979 and is designed to produce STABLEX in both slurry and solid form. In its slurried state, as in the U.K., STABLEX can be pumped to land reclamation sites to fill irregular extractive areas. The product, which looks like ready-mixed concrete, starts to set within 24 hours, and within 3 days is strong enough to walk on. Further strength development continues; after one month the product will support vehicular traffic. In its solid form the product is extruded after reaction to produce a pellet similar to aggregate.

Development in the U.S. has followed what has now begun to be regarded as a familiar pattern. Following a request from the State of Michigan for independent product testing, the National Sanitation Foundation at Ann Arbor, Michigan, a nonprofit organization of national and international repute, was engaged to carry out an independent test program using automotive wastes. The results of this testing [2] confirmed the environmentally safe nature of the product STABLEX.

This independent test work has been an essential prerequisite before discussing the project with local residents. The site chosen for the facility is 200 acres located in an area of outstanding natural beauty. On this 200-ac site, however, sand and gravel is extracted at some 800,000 ton/yr, and it is expected that a total of some 15 million tons of sand and gravel will be extracted over the next decade. The ability to follow the mining plan with a reclamation plan to restore the land to its former use represents an attractive proposition to the alternative, which is to leave the area as a scar to be inherited by future generations.

The acceptability of STABLEX as a land reclamation material can be demonstrated by the following tables. The properties of an ideal reclamation material have been described earlier, and it is interesting to note how demolition debris, generally accepted as the most suitable and most inert of reclamation material, matches up to the requirements. Table I illustrates the leaching results of demolition debris and road material compared to the proposed secondary drinking water standards when using the leaching procedure performed in accordance with the EPA Proposed Hazardous Waste Regulations Extraction Procedure. From Table I it can be seen that all these traditional materials fail the leaching test when compared to the EPA Proposed Secondary Drinking Water Maximum Limits.

Table II shows the leaching test results of STABLEX produced from automotive wastes compared to rock salt and the sand which is currently being extracted from the Groveland site. It can be seen quite clearly that the

Table I. Comparison of EPA Leaching[a] Results[b]

Parameter (mg/l)	Demolition Debris			Road Materials		EPA Proposed Secondary Drinking Water Max. Limits
	Concrete	Brick	Asphalt	Rock Salt	Fill Sand	
Chloride	76	63	77	31,000	32	250
Copper	<0.05	<0.05	<0.05	0.10	<0.05	1
Foaming Agents	NA[c]	NA	NA	NA	NA	0.5
Hydrogen Sulfide	NA	NA	NA	NA	NA	0.05
Iron	<0.05	<0.05	<0.05	<0.05	<0.05	0.3
Manganese	0.08	1.45	0.27	0.04	0.30	0.05
Sulfate	21.7	133	17.6	349	1.4	250
Total Dissolved Solids	1,410	400	306	50,500	582	500
Zinc	0.05	0.08	0.10	0.04	0.02	5

[a]Leaching procedure performed in accordance with the EPA Proposed Hazardous Waste Regulations Extraction Procedure [*Federal Register* (December 18, 1978) p. 58946].
[b]From the National Interim Primary Drinking Water Regulations [*Federal Register* (December 24, 1975) p. 59566].
[c]NA = not available.

Table II. EPA Leaching[a] Results of STABLEX

Parameter (mg/l)	STABLEX-Treated Automotive Industry Waste		Road Materials		Applicable EPA Maximum Drinking Water Limits[b]
	Waste Treatment Sludge	Zinc Bonderite Sludge	Rock Salt	Fill Sand	
Arsenic	<0.05	<0.05	<0.05	<0.05	0.05
Barium	<0.1	<0.1	<0.1	<0.1	1.0
Cadmium	<0.01	<0.01	<0.01	<0.01	0.01
Chromium (VI)	<0.05	<0.05	<0.05	<0.05	0.05
Lead	<0.05	<0.05	0.11	<0.05	0.05
Mercury	<0.002	<0.002	<0.002	<0.002	0.002
Selenium	<0.002	<0.002	<0.002	<0.002	0.01
Silver	<0.01	<0.01	<0.01	<0.01	0.05

[a]Leaching procedure performed in accordance with the EPA proposed Hazardous Waste Regulations Extraction Procedure [*Federal Register* (December 18, 1978) p. 58946].
[b]From the National Interim Primary Drinking Water Regulations [*Federal Register* (December 24, 1975) p. 59566].

STABLEX product is comparable in its results to the material being extracted, and in all cases meets the EPA Maximum Drinking Water Limits.

The method of land reclamation to be used at Groveland involves dividing the reclamation area up into cells of approximately four acres each. The cells will be reclaimed by pumping STABLEX into the cell and, after completely filling, covering the area with 3 ft of soil for return to agricultural use.

Provision is made throughout the whole of the reclamation area for three lakes which will act as recharge sources for the below-ground aquifers, as the STABLEX is extremely impermeable.

Because of the high level of industrial wastes arising in western Michigan, it is envisioned that using this technique the 200-acre site will be reclaimed over a period of 20-25 years, after which time the plant will be dismantled and moved away.

The reclamation of land with stabilized industrial waste is, therefore, providing a practical solution to the problems associated with the disposal of hazardous and toxic industrial wastes. No other method, commercially and technically proven, can provide such a safe and commercially attractive method of disposal as the SEALOSAFE service. The service is currently capable of handling around 70% of industrial wastes arising and, to date, over 22,000 different wastes falling into over 2000 substantially different categories have been treated successfully.

REFERENCES

1. U.S. Environmental Protection Agency. "Hazardous Waste Proposed Guidelines and Regulations and Proposal on Identification and Listing," 40 CFR Part 250, Part IV, December 18, 1978.
2. McClelland, N. I., H. B. Maring, T. I. McGowen and G. E. Bellen. "Leachate Testing of Hazardous Chemicals from Stabilized Automotive Wastes," Chapter 7, this volume.

CHAPTER 9

FIXATION OF HEAVY METALS FOR SECURE LANDFILL DISPOSAL

Mahendra D. Sandesara
Chemical Waste Management, Inc.
Calumet City, Illinois

INTRODUCTION

Spent acids from metal cleaning operations, etching solutions like ferric chloride, by-products of herbicide manufacturing, plating industry wastes and inorganic specialty chemical wastes all contain heavy metals such as arsenic, cadmium, chromium, copper, lead, nickel, zinc and others in varying degrees.

Inorganic metals, being elemental by nature, cannot be destroyed. Therefore, disposal of wastes containing heavy metals must be done in a manner to prevent these metals from reentering the environment in high levels, whereby the toxic effects of these metals will be exhibited. The only alternative is secure disposal. The only secure disposal method used today is landfill disposal in a secure site.

The primary factor in determining a site for secure landfill disposal is the geology of the site in question. The best landfill sites are located above natural clay bases, with the permeability of the clay being 1×10^{-8} cm/sec (0.0103 ft/year) or less. Other features of natural clay are the ability to withstand high temperatures and pressures; flexibility, which promotes a tight seal; and a natural alkalinity which prevents the leaching of nonfixed heavy metals (Figure 1). However, if a landfill with this type of geology is not available, solidified sludge containing heavy metals in a fixed (insoluble) state can be co-disposed with municipal refuse. This means that a landfill site with less than the best geology can accommodate these types of wastes without the fear of heavy metals leaching into the environment.

Municipal refuse, being mostly organic by nature, biodegrades in the landfill environment and generates carbon dioxide, methane and one or more organic acids, e.g., acetic acid, butyric acid, propionic acid, etc. If a fixed

Figure 1. Acid tanker unloading waste acid into a holding basin lined with clay and 3-in. quarry rocks.

(insoluble) heavy metal solidified sludge contains more than 5% excess alkalinity as calcium hydroxide, it will not be leached out by any of the above mentioned organic acids. Most landfill sites which co-dispose metal hydroxides have a pH of 5.0 or greater. As an example, if a fixed solidified sludge containing heavy metals has a free alkalinity of 15% as calcium hydroxide, to neutralize this sludge to a pH of 5.0 would require 16,050 gal of 1% acetic acid/yd^3 of solidified sludge disposed [1].

From the above data and computations, it can be concluded that the conditions necessary to produce an undesirable pH state in the landfill (one which will allow heavy metals to leach out) are virtually unobtainable. So, how do we create a fixed, heavy metal-containing, solidified sludge?

SOLIDIFICATION OF HEAVY METALS

Of the above mentioned heavy metal-bearing wastes, the spent acid group comprises the largest portion of this particular waste. A spent acid with a high heavy metal content makes the problem of disposal more hazardous than just disposal of the heavy metals alone. Spent acids are water-soluble, which makes almost all the heavy metals which they contain water-soluble. Also, these acids are very corrosive to body tissues, making handling and disposal more difficult. If this waste were to be disposed by secure landfilling alone, chances are great that the heavy metals, being water soluble, will leach out from the landfill into the environment in the near future, not to mention the hazard to which the landfill workers would be exposed. The majority of spent acids that carry heavy metals are spent pickling sulfuric and hydrochloric acids, nitric acid and chromic acid.

Another waste material which exists in large quantities, contains heavy metals and is a source of potential pollution is baghouse lime waste from quicklime manufacturing processes. This waste, which contains calcium and magnesium oxides and silicates, exists as a fine, dusty power with an alkalinity as calcium hydroxide of 50-65%. Disposal here is also a problem from an air pollution standpoint; the dust generated from dumping large quantities has both ingestion and inhalation toxicities. However, the waste can be transported easily and transferred on arrival at a disposal site by enclosed trucks specially designed to deliver fine powders (Figure 2).

Now we have two primary sources of possible heavy metal pollution. The unique aspect of these two waste materials is that when they are combined in the right proportions, the acids are neutralized and the heavy metals undergo fixation by creating insoluble metal hydroxides.

Steel mills in the Chicago area generate many thousands of gallons of spent acid from their pickling operations (Table I). This acid contains 5-15% acidity as sulfuric acid and between 2-5% reduced iron (Fe^{2+}).

At Chemical Waste Management, Inc., Calumet City, IL (approximately 20 miles south of Chicago), in excess of 850,000 gal of spent pickling sulfuric

Figure 2. Pneumatic tanker unloading waste baghouse lime into silo for later use in neutralization.

Table I. Typical Acids Received for Treatment and Secure Disposal

Sample	% Acidity	% Fe^{2+}	Heavy metals (mg/l)						
			As	Cd	Cr	Cu	Ni	Pb	Zn
1	10.2 (H_2SO_4)	3.6	1.0	1.8	0.1	1.8	2.5	14.2	89.0
2	4.1 (HCl)	0.1	1.2	10.5	8.0	3050	1720	2.9	230
3	13.6 (H_2SO_4)	2.8	1.0	2.6	0.1	1.9	3.1	12.1	78.0
4	8.8 (H_2SO_4)	4.9	1.0	1.3	0.1	2.7	2.3	9.6	180.0
5	9.6 (H_2SO_4)	0.0	2.2	0.8	21,500	8.9	10.3	19.3	75.0
6	1.2 (HNO_3)	0.0	0.1	1.1	2.8	1.8	3.5	0.1	15.0

Neutralized & Fixed Product Leachate Results

Samples	Excess Alkalinity as $Ca(OH)_2$	Heavy Metals (mg/l)						
		As	Cd	Cr	Cu	Ni	Pb	Zn
1-6 above when combined in holding pond	15.8%	0.01	0.01	0.01	0.83	0.42	0.01	0.97

acid are received per month for disposal along with over 250,000 gal of hydrochloric, chromic and, to a much lesser extent, nitric acid. These acids are all combined in two acid holding basins with a capacity of 350,000 gal each. The first transformation which takes place when the acids are combined is a change in the chromic acid, whereby chrome is reduced from the hexavalent to the trivalent state. The reaction is explained by the equation below:

$$H_2Cr_2O_7 + 6FeSO_4 + 6H_2SO_4 \rightarrow 3Fe_2(SO_4)_3 + Cr_2(SO_4)_3 + 7H_2O$$

Once chrome is reduced, neutralization with the baghouse lime (Figure 3) produces insoluble chromium hydroxide, which is nonleachable in secure landfill.

This neutralized product comes out as a slurry with 15% excess alkalinity as calcium hydroxide. On contact with air this slurry undergoes oxidation and solidifies within 45 minutes (Figure 4). The equations explaining the results are listed below:

$$H_2SO_4 + FeSO_4 + 2Ca(OH)_2 \rightarrow 2CaSO_4 + Fe(OH)_2 + 2H_2O \text{ (slurry)} \quad (1)$$

$$4Fe(OH)_2 + 2H_2O + O_2 \rightarrow 4Fe(OH)_3 \text{ (solid mass)} \quad (2)$$

The resultant product is easy to handle and is now ready for secure landfilling in a clay-lined site.

Secure landfilling is accomplished as follows:

A trench is designed to accept both solid and liquid waste. This trench is excavated into natural clay, lined with clay if the natural terrain is highly permeable, or artificially lined with an impermeable, manmade substance.

Figure 3. Acids being neutralized in the pug mill with waste baghouse lime.

Thus prepared, the trench, or cell, will accept liquid and solid wastes (Figure 5) in a ratio not to exceed 10 gal/yd^3 of solid refuse [2]. This ratio provides a "safe" level of moisture in the landfill, preventing oversaturation.

Upon completion of the cell, a final cover of natural clay or manmade impermeable material is placed over the trench and sealed. This is done to prevent rainwater from entering the cell and thus reducing the chances of possible leachate.

This process was taken one step further by adding a third waste material to the scheme, an arsenical salt by-product from a herbicide manufacturing process. This waste contained both soluble organic and inorganic arsenic, which needed to be detoxified before landfilling was possible. The detoxification process can be summarized as per the following equations:

For inorganic arsenic:

$$4As(ONa)_3 + 5FeSO_4 + 4Ca(OH)_2 + 5H_2SO_4 \rightarrow FeAs_2 + 2FeAsO_4 + 6Na_2SO_4 + 2Fe(OH)_3 + 4CaSO_4 + 6H_2O$$

For organic arsenic (in this case, cacodylic acid):

$$2\left(\begin{array}{c} OH \\ | \\ CH_3-As-CH_3 \\ \| \\ O \end{array}\right) \xrightarrow[\text{(acid media)}]{Ca(OH)_2} \begin{array}{c} CH_3 \qquad\qquad\quad CH_3 \\ | \qquad\qquad\qquad | \\ O{=}As-O-Ca-O-As{=}O \\ | \qquad\qquad\qquad | \\ CH_3 \qquad\qquad\quad CH_3 \end{array}$$

Figure 4. Neutralized acids are air-oxidized for solidification.

By combining the waste acid, baghouse lime and arsenical salt in the proper proportions, an insoluble form of arsenic was created without disrupting the existing acid neutralization, metal fixation process. This treated arsenical waste is now fit for secure landfill disposal, having been made 99.9+% insoluble. This process will also work for other solid, heavy metal-bearing waste, provided the waste is compatible with the existing system, i.e., pulverized solids rather than large chunks.

A U.S. patent for this fixation process has been issued to Chemical Waste

Figure 5. Co-disposal of neutralized and solidified acids at a secure site.

Management, Inc., due to its uniqueness, simplicity and effectiveness; a one-stage reaction using three waste materials, which when combined, results in a nontoxic, easily handled and suitable landfill material.

CONCLUSIONS

It has now been shown that the technology available today can be put into use to help alleviate the problem of toxic, heavy metal-bearing wastes. By conversion of these metals into insoluble compounds and disposal by secure landfill, these methods are both environmentally sound, as evidenced by the leachate results, and cost-effective due to the waste products involved in the fixation process. This then allows the supplier of this service to be competitive in the waste disposal field.

REFERENCES

1. Sandesara, M. "Process for Disposal of Arsenic Salts," U.S. Patent #4118243 (1978).
2. Miller, M. Illinois Environmental Protection Agency. Personal communication.

CHAPTER 10

THE METHODOLOGY USED IN DEVISING A NEW SOLUTION TO THE DISPOSAL OF HAZARDOUS WASTES USING URANIUM TAILINGS AS A CURRENT EXAMPLE

Robert L. Longfellow and David C. Hoffman

Dravo Lime Company
Pittsburgh, Pennsylvania

INTRODUCTION

Developing a new, viable solution for the disposal of inorganic slurry wastes requires considerable research and development (R&D) and a detailed understanding of the commercial impact from both a seller's and user's viewpoint. This chapter extends the technology that was specifically developed to resolve the slurry waste disposal problem resulting from the use of lime or limestone reagents in flue gas desulfurization (FGD). This technology involves the use of Dravo's Calcilox®* additive. Secondly, the discussion will describe the application of this technology to solve the problems of disposing of other inorganic slurry wastes. Overall, this chapter will summarize the development of Calcilox, briefly define stabilization and, most importantly, discuss the method used in applying this technique to various inorganic wastes.

DEVELOPMENT

In the early 1970s, Dravo performed extensive research on improving sulfur removal from coal-fired utility stack gases. Regulations were becoming very strict in reinforcing the concept of scrubbing: introducing an alkaline reagent into the stack gas stream so that the sulfur can be removed chemically. Extensive laboratory and pilot-plant work enabled Dravo to develop and patent a magnesium-containing lime scrubbing process: Thiosorbic®* lime

*Registered trademark of Dravo Corporation.

process. Thiosorbic lime has since received wide acceptance in the power industry as the best method for achieving greater than 85% SO_2 removal from sources burning high-sulfur (>3%) coal (Table I).

Throw-away flue gas desulfurization processes with lime and/or limestone scrubbing reagents produce another problem, i.e., the disposal of tremendous volumes of calcium-based waste slurry. These slurries are thixotropic (easily fluidizable), not readily handlable and not amenable to landfill. Concurrent with the development of the Thiosorbic lime scrubbing process, Dravo developed and patented a sludge treatment system using an additive called Calcilox. Calcilox is a dry, free-flowing, light-grey powder (Figure 1). When Calcilox is properly added to wet SO_2 scrubber sludges or selective inorganic slurries, it produces a stable, acceptable landfill material [1] which behaves like compacted soil, with definitive engineering properties that can be used.

DISCUSSION

The above briefly describes the impetus behind the development of Calcilox. Primarily, Calcilox is a specially water-granulated blast-furnace slag interground with various additives. The purpose of the water granulation is to trap large amounts of internal energy inside the granular particle by instantly cooling the molten slag from above its liquid temperature to well

Table I. Utilities Using the Thiosorbic Lime Process

Start-Up Date	Megawatts	Utility
Operational		
11/75	825	CAPCO–Bruce Mansfield #1
12/76	400	C&SOE–Conesville #5
10/77	825	CAPCO–Bruce Mansfield #2
1/78	410	Duquesne Light–Phillips
1/78	510	Duquesne Light–Elrama
4/78	400	C&SOE–Conesville #6
Contract Awarded		
3/79	625	Allegheny Power–Pleasants #1
12/79	250	Big Rivers Electric Coop.–Green #2
3/80	625	Allegheny Power–Pleasants #2
4/80	825	CAPCO–Bruce Mansfield #3
12/80	250	Big Rivers Electric Coop.
Letter of Intent Signed		
3/80	500	Kentucky Utility
7/81	600	Ohio Utility
	Total 7045	

Figure 1. Calcilox additive, a dry, free-flowing powder.

below its solid temperature. As a result of this trapped energy and chemical composition, this granulated material possesses cementitious properties when the material comes in contact with water. In the manufacturing process, the fineness of grind and the various additives to the basic slag material greatly influence stabilization rates and strengths.

Dravo Lime Company experience has shown that an economically viable Calcilox treatment system will require a Calcilox addition of 5-15% of the dry solids weight in the waste slurry. The actual chemical mechanism for stabilization is not straightforward and is beyond the scope of this chapter. The rate of Calcilox stabilization primarily depends on two variables: (1) the amount of Calcilox added, and (2) the concentration of inorganic waste solids. For example, if one adds a 10% Calcilox dose, one can expect a faster rate of stabilization than that achieved with a 5% addition to a higher percent solids waste slurry. The exact ratio of Calcilox additive to stabilization rate must be determined by laboratory testing (Figure 2).

Thus far, stabilization has been mentioned without much definition. Dravo Lime Company defines stabilization as a gain in unconfined compressive strength in the volume of settled solids or dewatered cake material. The unit of measurement used in this stabilization definition is ton/ft^2 unconfined compressive strength. The rate of stabilization is measured as the rate gain in unconfined compressive strength per day. Calcilox stabilization is somewhat slow in that significant strength gains (1-2 ton/ft^2) generally do not occur until the material has cured for 2-3 weeks for low-solids materials (25-35%). For high-solids materials (55-75%), centrifuge and vacuum disc filter cakes, adequate stabilization usually occurs after 3-5 days of curing.

In dealing with SO_2 scrubbing sludges, Calcilox is usually applied to a 30-35% solids thickener underflow. In this case, the Calcilox addition is usually in the 5-10% additive range and will result in unconfined compressive strengths of greater than 2.5 ton/ft^2 in 14-21 days after placement in temporary settling ponds or behind permanent impoundments. Of equal importance to the gain in unconfined compressive strength is the reduction in water permeability. Calcilox stabilization generally reduces the water permeability of FGD sludges by 1-2 orders of magnitude. Typical permeabilities

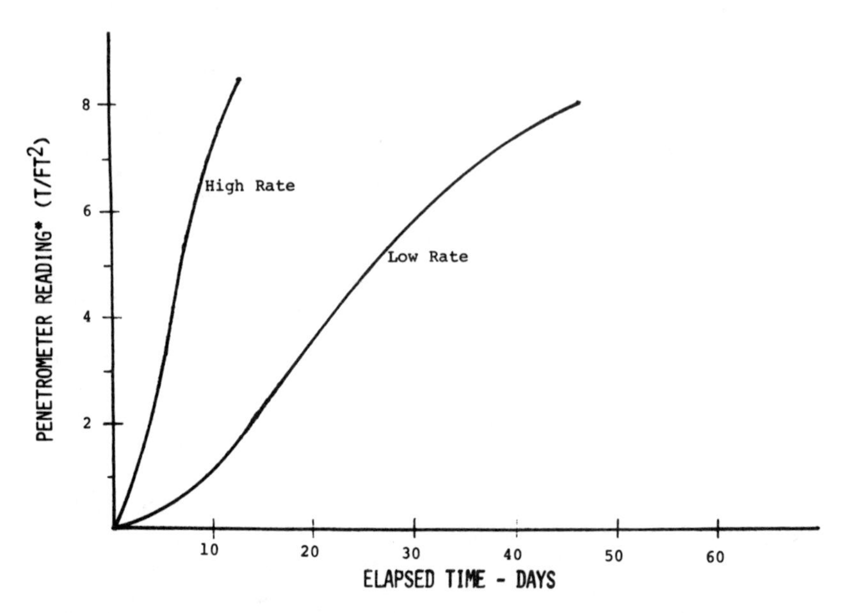

Figure 2. Example of Calcilox stabilization rate.

of Calcilox-stabilized FGD sludges will have a coefficient of permeability of 10^{-5}-10^{-7} cm/sec (1.2 in./yr). Thus, contamination of aquifer systems is greatly reduced with these low flowrates. A third important parameter that is improved with Calcilox stabilization is internal shear strength of the stabilized mass. Generally speaking, Calcilox stabilization will impart an internal cohesive strength greater than 10 psi on an initial 30% FGD solids slurry with a 10% Calcilox addition. The actual internal shear strength will be a function of initial solids concentration, amount of Calcilox additive, stabilization time and the initial concentration of fly ash in the untreated FGD waste sludge.

Three Methods of Treatment and Disposal (Figure 3)

Local conditions and topography usually control the method chosen for treating, transporting and disposing of sludges. The proportion of Calcilox hardening agent and the stabilization time required both depend on the chemical and physical characteristics of the waste material being stabilized and on the disposal system being used. Laboratory analysis and field data

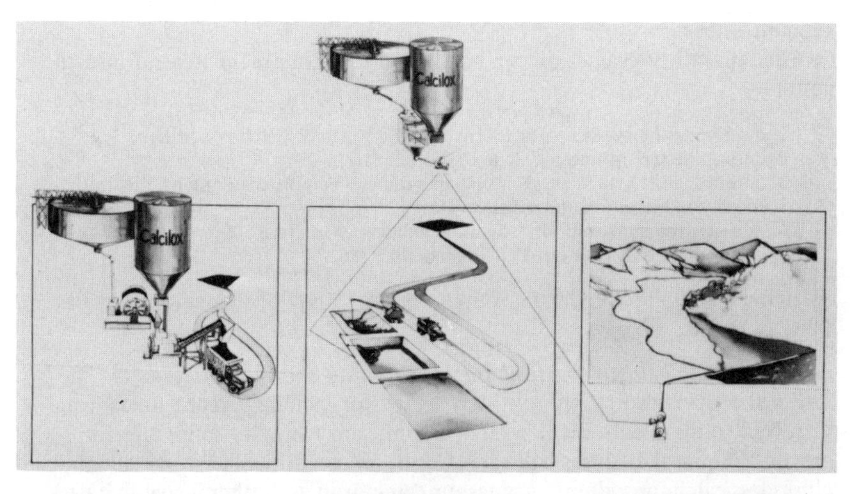

1. The sludge or slurry may be mechanically dewatered to a semi-solid state. The stabilizing agent is then added and the mixture is cured temporarily and then delivered into a permanent landfill site.

2. Sludge or slurries are treated with the Calcilox additive then held in temporary curing ponds (usually from 5 to 20 days depending on sludge characteristics) until they acquire the necessary hardness. The stable material is then excavated and transported with ordinary earth moving equipment to a final disposal site.

3. The sludge may be delivered to a permanent impoundment site in a slurry form. Addition of Calcilox to the slurry (or thickener underflow) may be done at the point of origin or just prior to disposal. The solids of the Calcilox treated slurry placed into the impoundment will settle and stabilize under water. The supernatant (water) can then be returned for use in the plant thus conserving water.

Calcilox and Thiosorbic are registered trademarks of Dravo Corporation.

Figure 3. Three methods of treatment and disposal.

accumulated over a five-year period by Dravo Lime Company are used to determine the most economical proportion of additive and to select the disposal method for optimum performance.

LABORATORY APPROACH TO NEW WASTE APPLICATIONS

The disposal of all types of waste materials via landfills is becoming more and more difficult due to environmental considerations. Thus, in devising any laboratory test to determine the feasibility of applying Calcilox stabilization techniques, one must keep in mind that the program should simulate the actual field conditions as closely as possible. Two of the prime parameters that must be investigated in all laboratory programs are the leachability of the material and the immediate and final handling characteristics as placed in the landfill. Of these two parameters, leachability is probably one of the most difficult laboratory tests to perform to obtain meaningful results that can be equated to the actual field conditions.

The first important step in any laboratory program is to obtain representative samples of the material. Based on experience, most samples are taken either from a thickener underflow, an excavated settled solids material from a settling pond or a dewatered cake material taken from a centrifuge or vacuum filter.

The laboratory evaluation can be divided into four main areas of investigation:

1. determine the necessity for pretreating the waste or Calcilox additive;
2. determine the optimum Calcilox addition rate;
3. determine the effects of such system variables as moisture content, pH, initial solids concentration and settling rate;
4. measure the improved physical and chemical characteristics imparted to the waste slurry as a result of Calcilox stabilization.

Items 2 and 3 are the heart of the program. Some of the major points in these items are:

1. *Physical/chemical waste characterization of the untreated waste.* The physical properties that are routinely tested for include percent solids, total dissolved solids (TDS), settling rate (if applicable) and dry solids specific gravity. Chemical analyses will vary. Common constituents are pH, alkalinity, chlorides, calcium, sodium, magnesium, zinc, iron and other heavy metals.

2. *Stabilization testing.* Calcilox is added in varying doses to the waste material to determine the optimum rate of stabilization to meet disposal requirements. Based on Dravo Lime Company's experience, a material attaining an unconfined compressive strength of 2-2.5 ton/ft^2 is adequate for immediate handling and/or placement and compaction in a landfill.

3. *Unconfined compression testing.* This parameter is an accurate indication of the actual stabilized strength of the material, neglecting the effects of containers or a large mass of material. Extensions of this type of testing can

yield other soil mechanics data such as shear and cohesive strength. Because of experience, extensive testing along these lines is not normally performed.

4. *Permeability tests.* This test, coupled with the unconfined compressive results, is very important in landfill applications. The actual test is quite simple. A carefully sized specimen of stabilized material is placed in a device where waste supernatant is allowed to permeate through the specimen. Calcuations are performed depending upon the geometry of the test apparatus to yield a permeability coefficient for either vertical or horizontal flow. Typical values of Calcilox-stabilized materials range from 10^{-5} to 10^{-7} cm/sec. For realistic data on leachate quality, the permeate must be gathered from these test specimens and analyzed for each particular parameter. Since the coefficients are quite low, leachate collection can be difficult.

When all the results from the above-mentioned testing are gathered, a realistic waste disposal concept can then be designed. If the results of these tests are not acceptable to the client and/or regulatory agencies, a small field demonstration can usually be conducted and the questionable areas can then be examined on a larger scale.

To get a better feeling for what has just been discussed, here is an example of how this type of approach has solved one of the major problems in the coal industry today, i.e., the disposal of the fine coal refuse generated in the preparation of coal. Initial developmental work consisted of visiting coal mines all over the U.S. and collecting samples of the fine coal refuse that is discharged into settling ponds, behind permanent impoundments or processed into dewatered filter cakes or centrifuge cakes. Dozens of these samples were analyzed, tested and stabilized with Calcilox. From this sampling, a company was selected to participate in a demonstration of Calcilox stabilization techniques.

FIELD DEMONSTRATION

This test was conducted in batches, using large concrete mix trucks. These concrete mix trucks received the thickener underflow and mixed the material while the Calcilox was being added to the batch. After the waste slurry and Calcilox were thoroughly agitated to disperse uniformly the Calcilox throughout the slurry, the concrete mix trucks discharged their contents into a below-grade stabilization pond approximately 28 X 40 X 8 ft deep. This pond received approximately 14 truckloads of slurry amounting to ~450,000 lb. The average solids concentration was 45% and received an average dose of approximately 15% by weight of Calcilox. Monitoring of the pond was accomplished in the laboratory by taking individual truckload samples and measuring their rate of penetration resistance. The average results of these individual truckloads indicated that the pond material would achieve 4 ton/ft^2 penetration resistance within a 2-week period. More sophisticated unconfined compression results indicated that the material actually possessed a 3-ton/ft^2 unconfined compressive strength at the end of this 2-week period.

The pond was allowed to remain undistrubed for 30 days, in which samples of the pond were taken and excavation of the stabilized material was accomplished with a D-7 bulldozer. Laboratory results of the excavated material indicated that it had an in-place bulk density of 81 lb/ft^3 with a solids content of 58%. Further laboratory testing indicated that the unconfined compressive strength amounted to slightly over 8 ton/ft^2 (Table II). As a result of this field demonstration, laboratory results became easily and visibly equatable to a physical reality. Today, there are three full-scale Calcilox systems treating in excess of 100 ton/hr of fine coal refuse with this interim pond approach. Several other large systems are in the design stage using a permanent impoundment approach that will take into consideration the in-place strengths of the Calcilox-stabilized fine coal refuse (see Figures 4-6).

URANIUM MILL TAILINGS

The uranium industry produces over 7 million tons of tailings annually in the United States and currently has 22 inactive tailing ponds that need attention. An in-depth study by the Nuclear Regulatory Commission (NRC) has determined that the handling and placement for disposal of uranium tailings is in need of new approaches and technology. The NRC and the industry are concerned about groundwater contamination, air pollution from blowing dust and radiation pollution. They are concerned about the integrity of the disposal site today and for generations to come.

Using the approach of laboratory testing followed-up by field demonstration work, Dravo Lime Company is currently involved in testing uranium mill tailings stabilization with Calcilox additive. One of the major concerns in processing uranium ore is the disposal of the waste tailings. Not only do these tailings pose handling and transportation problems due to their thixotropic nature, the more pressing area is the emission of radon gas from the tailings. This radon emission problem is most difficult because it is a gas and more difficult to control and handle than leachates or solid waste material.

Table II. Fine Coal Refuse Field Demonstration–Unconfined Compressive Strengths

Location	Strength (ton/ft^2)
Undisturbed Pond Samples	
30 days cure	7–8
Mix Truck Samples, as placed in pond	
15 days cure	3.0
30 days cure	7.6
90 days cure	9.8

Figure 4. Excavating untreated settled fine coal refuse.

Figure 5. Excavating Calcilox additive-stabilized fine coal refuse.

Figure 6. Loading Calcilox additive-stabilized fine coal refuse.

The radon gas decay pattern for radon is depicted in Figure 7. Only the four daughters, RaA(Po-218), RaB(Pb-214), RaC(Bi-214) and RaC(Po-214), have significance as far as rapid release of alpha energy is concerned. Excessive exposure can cause cancer. However, if these gases are retained within the tailings as they are produced, for only a few hours, they will have decayed further into harmless forms.

Laboratory tests dealing with the uranium tailings have been designed to correspond to a waste disposal program which currently encompasses the relocation of the tailings from their present disposal area to a more suitable location. The primary objective of the stabilization program is to reduce the radon emissions and to improve the handling and compaction of the re-excavated tailings. No effort has been made to recover uranium that may be contained within the tailings deposit.

Dravo Lime Company's initial testing and economic analyses indicated that the cost of treating 100% of the uranium mill tailings with Calcilox was too high. The industry could and would not accept the necessary increase in cost. Alternative disposal concepts were conceived as a means for reducing the Calcilox consumption while still achieving stabilization. These concepts, which require laboratory and field testing to develop and demonstrate their effectiveness, are as follows.

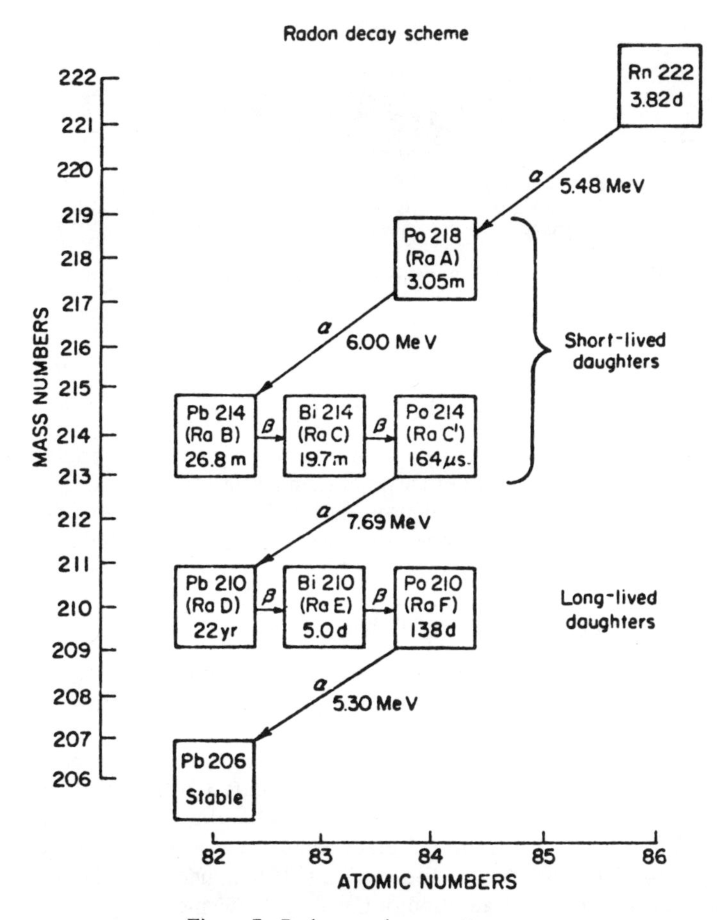

Figure 7. Radon gas decay pattern.

Pond Sealing and Capping

One concept is to treat a small percentage of the tailings and create an artificial seal of 3-4 ft thickness for the pond bottom. After filling the sealed pond area with untreated tailings, the settled solids would then be capped with a layer of treated tailings (Figure 8). This capping technique may be developed and tested for use in covering inactive tailing ponds as well. The Department of Energy (DOE) has as a current objective the development of a technique so that 22 inactive ponds may be covered to reduce radon gas emissions to acceptable limits, i.e., twice background levels.

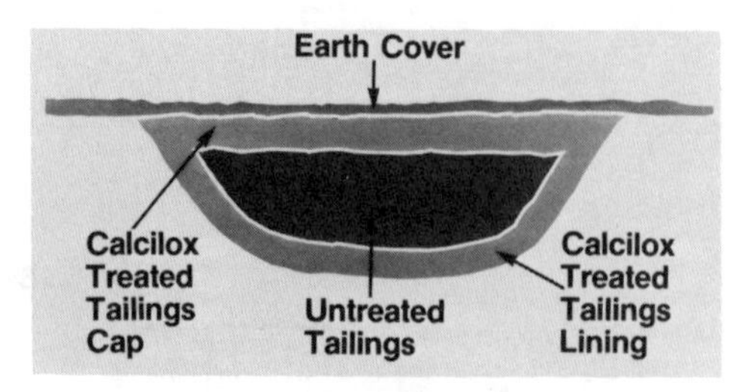

Figure 8. Pond sealing and capping.

Treating Only the Fine Portion of the Tailings

Some plants generate two sizes of tailings; one is a sandy material (~70%) that can be disposed of without treatment, and a slimes portion (~30%) which can be stabilized using Calcilox additive (Figure 9). If all tailings streams can be split in this manner, the cost of the Calcilox stabilizing technique becomes acceptable.

Low Calcilox Addition to All the Tailings

An alternative theory has been developed that required full-scale field testing. If Calcilox is added to the tailings stream containing both sandy material and slimes, the rate of addition may be very low if Calcilox would tend to remain with the slimes portion. When discharged into a pond, the tailings stream experiences a natural size segregation (Figure 10). The sandy portion drops out immediately due to natural settling while the slimes flow on to the far side of the pond. It is expected that Calcilox will carry on with the slimes portion since its consistency is also very fine. If this theory is proven correct, rather than a normal 10% addition of Calcilox stabilizing agent, only a 3-5% addition would be necessary in the total stream to achieve an actual addition rate to the slimes of approximately 10%.

Laboratory tests have already proven the Calcilox additive as a potentially attractive solution to the problem of stabilization of uranium tailings (data

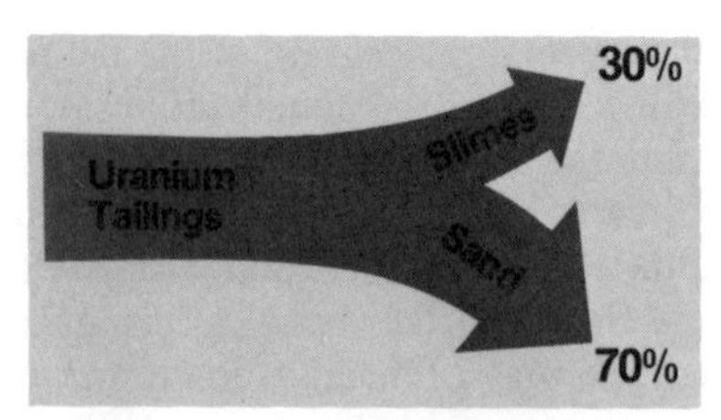

Figure 9. Treating fine portion only.

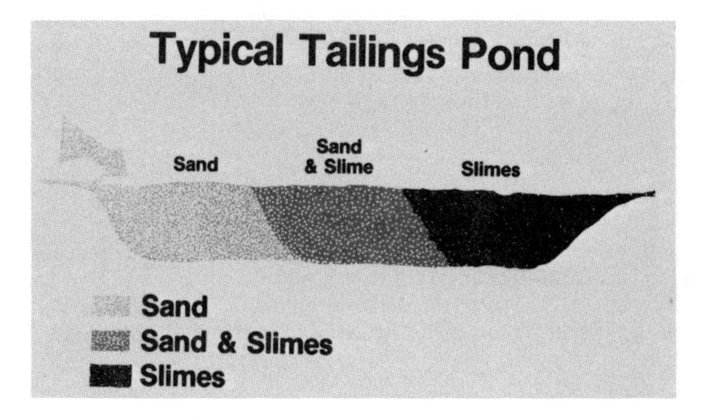

Figure 10. Sands drop out.

not yet ready for publication). Tests indicate that permeabilities are reduced to the 10^{-6}-cm/sec range, load-bearing capabilities are increased to ~4-5 tons/ft^2, and (unique to these wastes) radon gas emissions are significantly reduced. Currently, both laboratory and field tests are under way. Dravo Lime Company is working as a subcontractor of Ford, Bacon and Davis. They have a prime contract with the Department of Energy to develop the best and least expensive technique for recovering inactive uranium tailing ponds to reduce the radon gas emissions to acceptable levels. With a successful completion of these tests the next step will be to field-test the disposal concepts outlined above.

OTHER APPLICATIONS

From Dravo's R&D it has been found that Calcilox stabilization techniques are most applicable to various inorganic wastes. The economic solution to many of these inorganic disposal problems is greatly affected by various characteristics of the waste slurry. In particular, the percent solids, particle size, specific gravity and settling rate of the solids are most influential in achieving satisfactory stabilization of the material. Current research is conducted in fluidized bed combustion wastes and dredging wastes with such pollutants as polychlorinated biphenyls. By following the procedures as discussed in this chapter, Calcilox stabilization techniques may be applied successfully to these new areas and to other types of inorganic wastes. As a final note, the real criteria for any practical waste disposal approach are design objectives, which must be matched with the treatment method and ultimate disposal mode. The best way to accomplish this is by a logical sequence of laboratory testing, preliminary engineering to estimate system economics and, as necessary, pilot programs to clarify borderline decisions. If this approach can be employed as discussed, slurry waste disposal will be technically sound with optimum economics.

REFERENCE

1. Labovitz, C., and D. C. Hoffman. "Effective Disposal of Fine Coal Refuse and Flue Gas Desulfurization Slurries Using Calcilox® Additive Stabilization Technique," in *Toxic and Hazardous Waste Disposal, Vol. 1*, R. B. Pojasek, Ed. (Ann Arbor, MI. Ann Arbor Science Publishers, Inc., 1979), pp. 93-102.

CHAPTER 11

USE OF PLASTIC SOLIDIFICATION AGENTS WITH THE EXTRUDER-EVAPORATOR VOLUME REDUCTION AND SOLIDIFICATION (VRS) SYSTEM

Richard D. Doyle

Werner & Pfleiderer Corporation
Waldwick, New Jersey

INTRODUCTION

Hazardous wastes, the residues of normal industrial processes, are produced at an ever-increasing rate. The U.S. Environmental Protection Agency (EPA) estimates that 30 million ton/yr are generated. The production of these wastes cannot be prevented, but must be controlled by industry. This presents industry with a problem in meeting its obligation to its investors and customers to produce an end product as economically as possible.

Before recent promulgations by regulatory agencies making the originator/producer responsible and liable for its wastes, even after disposal, there was little incentive for waste treatment. Although the moral obligation to protect the public and the environment was important, the cost of waste treatment and its effect on product competitiveness far outweighed these considerations. However, recent EPA studies have shown that the cost incentive is now present when ultimate responsibility must be assumed by the producer. Wastes that would have cost $100,000 to treat properly, as produced, actually cost over $2,000,000 to recover and treat several years after production. This 20:1 cost differential, which will widen with inflation, provides industry with the cost incentive necessary for use of the best available technology (BAT) for treating their wastes. Enforcement of the regulations by EPA and other agencies will ensure protection of the public and the environment.

Industry has acknowledged the problem and has spent billions of dollars to develop waste treatment processes incorporating BAT. This is a very complex problem, and increases with every new product and subsequent waste stream. There is no universal waste treatment or immobilization process that will handle all the variations of waste produced. However, for most liquid wastes, the concentration of the hazardous components and their immobilization in a binder is imperative if we are to avoid being drowned in this waste. This concentration step also provides the added benefits of cleaning the solvent for reuse within the plant and reducing the final volume of waste for ultimate disposal.

Many processes have been developed for waste treatment with most aimed at a specific waste stream with known chemical characteristics. The volume reduction and solidification (VRS) system developed by Werner and Pfleiderer Corporation (WPC) was aimed at the need for a process to handle a variety of chemical properties associated with the relatively low-volume, highly toxic wastes such as arsenic, heavy metals or electroplating wastes.

The VRS system will produce a solidified end product with the hazardous constituents, regardless of their chemical characteristics, immobilized in a plastic binder. The flexibility and capability of this system to handle a wide variety of liquid and dry waste feed streams in conjunction with most plastic binders (e.g., asphalt, polyethylene, polypropylene, ureaformaldehyde and vinyl polyesters), means that industry has state-of-the-art technology available for the treatment of their wastes. In addition to having this flexibility, the VRS system will reduce the total volume of wastes produced, thus reducing ultimate disposal costs and problems.

The VRS system is a simple mechanical process where the extruder-evaporator in one step evaporates all unwanted solvent from the waste while homogenizing the waste constituents with the liquid plastic binder for discharge into a container. Once this waste-plastic mixture has cooled, a significant volume-reduction effect is realized in an end product that meets even the most stringent requirements. In addition to this, the system returns the clean solvent to the plant for reuse, thereby reducing plant makeup requirements. The end product is finely dispersed waste salts in a solidified plastic matrix.

The VRS system is designed to process liquid and solid wastes generated from industrial processes. This waste stabilization and solidification concept was designed and developed over the past 15 years by Werner & Pfleiderer, an affiliate of WPC, which is a leading supplier of compounding extruders for the plastic industry in Europe. WPC has applied this technology to the processing of liquid and solid wastes generated by industry, including the nuclear power industry. Application to waste handling is based on well-proven components.

Over 2000 extruder-evaporators are operating worldwide in the chemical, plastics, food and nuclear industries. In hazardous waste stabilization and solidification applications these systems have amassed a total operating history of over 50 unit operating years. This operating experience was obtained with minimal routine maintenance and no exposure of personnel to hazardous conditions during operation or maintenance.

The WPC-VRS system provides the producers of hazardous wastes with state-of-the-art technology in a system with proven reliability and flexibility.

When this system is selected and incorporated in an actual processing plant, the operator will have at his hand the many advantages of the WPC-VRS system including:

1. removal and recycle of all unwanted solvent in the waste, and subsequent stabilization and solidification of only the hazardous constituents;
2. volume reduction of the end product (up to 80% reduction compared to other processes), thereby reducing disposal space requirements (this is especially important when waste will be stored onsite);
3. processing capability for handling the wide variety of wastes generated by industry, thus allowing feeds from many different sources while maintaining assurance of solidification;
4. flexibility in selection of binders to meet the needs of industry and the varying chemical characteristics of their waste streams;
5. ease of operation–the extruder-evaporator evaporates all solvents and disperses the remaining waste solids into a plastic matrix in a single step, while the end product naturally solidifies in its container upon cooling;
6. use of batch or continuous modes of operation with demonstrated reliability, thus reducing downtime, maintenance requirements and operator attention;
7. lower operating costs, compared to nonvolume-reduction methods of disposal, due to reduced volume and weight shipped offsite, lower material costs for binder and extended life of existing permanent disposal facilities;
8. reliability–the extruder-evaporator has been proven in plastics compounding and hazardous waste processing applications to have an availability of 98% or greater; and
9. operator safety is assured by dust-free operation near ambient pressure, thereby eliminating the potential for in-process accidents, leaks and spills.

These benefits will be achieved by all users of the WPC-VRS system, which provides industry with a single system for treatment of its hard-to-handle, and ever-changing hazardous wastes.

VRS SYSTEM DESCRIPTION

Process

The WPC-VRS system is a one-step volume-reduction and solidification process. It employs an extruder-evaporator for removal of free and combined water, mixing the wastes with a plastic binder, and dispersing the waste salts in a homogeneous matrix. The system completely stabilizes and solidifies hazardous wastes of all types and characteristics. A full description of the process, research and development (R&D), and other data are provided in Doyle [1].

The extruder-evaporator is the major component of the WPC-VRS system. However, the complete system includes collection and feed systems; processing, discharge and containment; container handling systems; and remote-control instrumentation. The system description that follows will reference an asphalt system, as this is the most widely used binder. However, use of other plastic binders would be comparable.

Feed Systems

Wastes of all types (slurries, evaporator concentrates, ion-exchange resins, chemical drains, sludges, manufacturing residues, incinerator ash, etc.) are handled by separate mixing and pretreatment components. The homogeneously mixed wastes are then accurately metered into the inlet barrel of the extruder-evaporator along with a controlled flow of plastic binder. The ratio of binder to waste is predetermined to maintain an end product of 50% waste salts and 50% binder, by weight.

Binder Feed (Asphalt)

An asphalt storage tank is located in an easily accessible area, near the solidification system and truck delivery access. The tank is sized to accommodate a bulk truck shipment of hot liquid asphalt to optimize delivery, storage and operational requirements. The hot asphalt remains liquid by maintaining its temperature during storage above 250°F.

The asphalt feed system, including transfer and metering pumps, strainers, heat tracing, instruments, and controls is designed for operation from a main control panel.

Wastes

The WPC-VRS system includes a complete feed system for each type of waste. Each feed system includes the required tanks, mixers, transfer pumps, metering pumps, feed conveyors, piping, valving, instruments and controls. These feed systems are interlocked to prevent wastes from being fed to the unit without the extruder-evaporator screws turning, containers in position, or asphalt flowing.

Processing

The processing operations (volume-reduction and mixing) are accomplished in a motor-driven, twin-screw extruder-evaporator. Twin co-rotating screws are used to mix and convey the waste product and immobilizing agent along the length of the machine. The housings or barrel sections in which the screws rotate are heated to evaporate free and entrained solvent from the mixture. The solvent removed from the feed stream is condensed in the steam (vent) domes, and recycled to the plant.

The asphalt and waste are fed to the extruder-evaporator in the first barrel, with the waste slightly downstream of the asphalt. The waste salts and the asphalt are kneaded into a homogeneous mixture of microscopic particles, each coated with asphalt. Entrained liquid is evaporated up through the steam domes and recycled back to the plant. The homogeneous mixture of asphalt and waste is then discharged into a container for ultimate disposal.

The product from the extruder-evaporator can be placed in almost any type of container, e.g., 55-gal drums, 50-ft^3 liners or flat forms for foundation work. The ultimate disposal of the filled container is the deciding factor on the configuration and type of container to be used.

The WPC-VRS system is a completely enclosed system and, therefore, is compatible with remote operation and isolation that may be required for handling some highly toxic material. A typical system for this type of operation is used for radioactive wastes and is described in the following paragraphs. However, it should be remembered that where the less-hazardous materials are handled, less-complex methods may be desirable and more economical. For highly toxic or radioactive wastes, a separate, isolated filling chamber is located at the end of the extruder and includes the extruder-evaporator discharge barrel plus a turntable for positioning drums under the discharge barrel. The turntable holds 4-8 drums, and is indexed to accurately position each drum in turn under the discharge barrel during filling.

For complete remote operation, an overhead crane is used to place empty drums on the turntable and transfer the filled ones. Filled drums are normally stored in the filling chamber for cooling before being moved to interim storage or shipment to a disposal site.

For extremely hazardous materials (e.g., radioactive wastes), a remotely operated drum capper and monitoring station is provided to cap the filled drums and monitor toxic material external contamination levels. Should a drum be contaminated, it is placed in the cleaning station. Steam and hot water are used as the cleaning agents. Other container configurations can be handled in a similar manner.

The WPC-VRS system is designed and engineered with an appropriate instrumentation and control package to match the needs of the process. In the simplest systems, all operations are manually initiated from local racks. In systems where hazardous materials are handled, or where minimal operator attention is desired, a totally automated system is provided. Any degree or combination of these two extremes can easily be integrated to be compatible with a specific process.

BINDER CHARACTERISTICS AND SELECTION

Recent trends in hazardous waste stabilization/solidification have been toward plastic solidification agents. WPC has been involved in the North American plastics processing industry since 1959. The successful use and reliability of the extruder-evaporator in this industry has included development of processes and systems for the manufacture and use of various types of plastic materials, some of which have application to hazardous waste stabilization/solidification.

Plastics can be separated into two general classifications: thermoplastic or thermoset materials. WPC has used thermoplastic materials in most applications to date, with asphalt and polyethylene being used most extensively. The paragraphs below describe why thermoplastics are presently considered superior to thermosets (such as ureaformaldehyde and polyesters) as hazardous waste stabilization/solidification agents. However, due to the flexibility of the extruder-evaporator, use of thermosets with their required multiinjection points for carriers, promoters, hardeners and catalysts, is practical, and will be utilized when benefits accrue.

Thermoplastic Materials

These materials can most easily be described as materials whose behavior is analogous to paraffin. When heated they become liquid and when cooled, they are solid. This heating and cooling process can be repeated as often as required. Materials which exhibit these thermoplastics properties include asphalt, polyethylene, polypropylene and nylon.

Thermoset Materials

Thermoset materials, on the other hand, behave in a manner analogous to an egg. When they are heated, a chemical reaction is initiated which hardens the material. Once a thermoset is solid, it will remain solid even if subsequently heated. With all thermoset materials, mixing, heating and/or action of a catalyst initiates a cross-linking reaction. This reaction is influenced by such parameters as pH, water content, ionic constituents in feed streams, temperature and homogeneous dispersion of catalyst in the mixture. Therefore the thermosetting reaction is somewhat unpredictable if it is not operated under totally controlled conditions. Materials that exhibit the properties of thermoset materials include ureaformaldehyde, polyesters, phenolics, and melamine.

In many ways cement is analogous to a thermoset plastic material. With cement a chemical reaction is initiated that is somewhat unpredictable, difficult to control and exothermic. This reaction is extremely sensitive to feed stream chemistry resulting in the risk of free water or nonsolidification of the end product if conditions are not properly controlled.

WPC has extruders operating in plastics applications with all of the plastic materials listed above. Over 2000 Werner & Pfleiderer plastics processing facilities are operating worldwide, and the knowledge and experience gained both at WPC and their affiliate, Werner & Pfleiderer, Stuttgart, West Germany, have been used in the evaluation of solidification agents for radioactive waste applications.

Of the many plastic solidification agents that are compatible with WPC equipment, the thermoplastic material asphalt has been found to have the best combination of economic, chemical and physical characteristics. This evaluation is based on a comparison of such features as economics, elimination of free water, assurance of a solidified product independent of process parameters, corrosion of end-product containers and compatibility with all types of wastes. In addition, asphalt has been shown to be a high-integrity barrier to environmental release of contaminants by its proven low leachability, high freeze-thaw resistance, inertness to bacterial attack and long-term stability.

The major component of the WPC-VRS system is the extruder-evaporator.

This highly versatile component is in extensive use in a variety of chemical, nuclear and plastics processing applications. WPC presently recommends the use of an asphalt binder based on its low cost, compatibility with the extruder-evaporator and the environmental acceptability of asphalted end products. However, other plastic binders, including polyesters, can be employed in the WPC-VRS system now, or at a later date, if considered desirable due to better compatibility or development of new materials.

OPERATING EXPERIENCE

The WPC-VRS system is one of the most extensively researched liquid waste solidification processes in the world. This research is well documented, and the European users claim the asphalt stabilization/solidification process is superior to the process previously used (cement, cement silicate mixtures and ureaformaldehyde) for a variety of reasons, i.e., volume reduction, cost savings, compatibility with waste feeds and superior end product. Application of this system to the fixation and encapulation of arsenic-laden wastes was tested by JBF Scientific, Wilmington, MA, under EPA Contract No. 68-03-2503. Analysis showed that there are great differences among the various processes. In general, the asphalt-stabilized/solidified samples showed a leach rate approximately two orders of magnitude below that of the next best method. Preliminary results of this study were presented at the Hazardous Waste Land Disposal Symposium in San Antonio, TX, March 8, 1978 [2].

A considerable portion of the research was done in conjunction with the nuclear industry [3,4]. Although the major hazard was considered the radioactive content of these wastes, the chemical composition and characteristics of the wastes handled are indicative of the hazardous wastes produced by other industries. Therefore, the process treatment technology that was developed is directly applicable to hazardous wastes of similar chemical characteristics regardless of their source. The operating experience and type of wastes handled at these facilities are summarized in Table I.

JBF Scientific and recent WPC testing has shown that the data developed for radioactive chemical species will be similar to those for handling of non-radioactive wastes. Future development with similar chemical constituents will be accomplished as specific problem areas and needs are recognized to determine applicability of the WPC-VRS system.

ULTIMATE DISPOSAL

The WPC-VRS system provides many advantages for the ultimate disposal of hazardous wastes when compared to other currently used methods of solidification/stabilization. Each of these advantages are available and can be realized, depending on specific operating requirements.

Table I. WPC-VRS System Hazardous Waste Operating Experience

Company	Location	Number of Units	Startup Date(s)	Materials Treated
Commissariat a l'Energie Atomique (CEA)	Marcoule, France	2	1965	Diatomaceous earth, ion exchange resins, nickel ferricyanide, barium sulfate, borates, sulfates, sodium nitrate, calcium nitrate, mercuric nitrate and aluminum hydroxide
Commissariat a l'Energie Atomique (CEA)	Cadarache, France	1	1969	
Karlsruhe Nuclear Center	Karlsruhe, West Germany	2	1972/73	Spent ion exchange resin, boric acid concentrates, sodium sulfate and filter sludges
PZEM	Borssele, Netherlands	1	1974	Boric acid, ion exchange resins and liquid waste
Comision Nacionale de Energia Atomica at Lima	Buenos Aires, Argentina	1	1974	Boric acid, ion exchange resins and sodium sulfate
European Company for the Chemical Processing of Irradiated Fuels (Eurochemic)	Mol, Belgium	1	1976	Sodium nitrate, sodium nitrate ferricyanides, barium sulfate, ammonium nitrate, sodium aluminate, sodium hydroxide and aluminum nitrate
Atomic Energy of Canada Limited (AECL)	Chalk River, Ontario, Canada	1	1976	Phosphates of iron, calcium and magnesium, calcium carbonate, magnesium carbonate, incinerator ash and sodium sulfate
EPPLE	Stuttgart, West Germany	1	1976	Oil residues
Werner & Pfleiderer Corporation[a] Development Laboratory	Waldwick, New Jersey	Various Test Facilities	Various	Test programs have included sodium sulfate, boric acid, bead resins, powdered resin, filter sludge, liquid waste (10% solids), decontamination solution (10% solids), sodium perborate, EDTA, hydrazine, ferric oxide, sodium hydroxide, sodium chloride, diatomaceous earth, phosphoric acid, arsenic trioxide, organic arsenicals electroplating (heavy metals) and filter aid.

[a]Operation of this facility is on an intermittent basis to support active programs in hazardous waste handling.

Volume Reduction

Total volume of waste to be ultimately disposed is reduced by a factor of 5-20 depending on the material to be stabilized/solidified. This will be beneficial in that it will effectively increase the capacity of in-plant storage and commercial disposal areas.

Operational Impact

The system is a simple, one-step, self-contained unit that can be totally enclosed, thereby allowing processing of even the most hazardous materials without exposing personnel to the dangers. No obnoxious or detrimental effluents are released during processing. In addition, energy requirements are minimal and the solvent can be returned to the plant for reuse, thereby reducing make-up requirements.

Transportation

Due to the large reduction in end product volume, transportation to the ultimate disposal site is minimized. This reduces the total number of shipments and energy resources required to transport waste to the site. With reduced shipments, there is also a significant reduction in the potential for an accident to occur that could release hazardous materials to the environment.

Disposal

The solidified wastes produced by the WPC-VRS system are a homogeneous mix of waste salts and asphalt. The solidified wastes are in the form of a monolithic solid with no free water, which meets regulatory requirements. Solidified wastes of this type have been subjected to extensive testing by others.

Solidified wastes from all systems are subjected to a variety of conditions and handling. The operation and testing of the use of the WPC-VRS system, and asphalt as a binder have shown that this system produces a superior end product.

Of concern to most individuals is the stability of the end product in the ground which in most cases is evaluated by determining the leach rate of the solidified waste salts. The leach rates of the asphalt waste salts mix is on the order of 10^{-4}-10^{-5} g/cm^2/day, which is approximately two orders of magnitude better than other currently used solidifying agents with the same waste concentrations. Determination of leach rates of samples of specific wastes encapsulated in asphalt is continuing. Leach rates using other plastic binders need to be evaluated on a case-by-case basis.

In recent tests a very low rate of leachability of radioisotopes in demineralized water was found. The results, using the International Atomic Energy Agency (IAEA) standard leach test procedure, are shown in Table II.

Other tests of leach rates with asphalt and salts microdispersed throughout show that the leach rate depends on the weight-percentage of salts present

Table II. Leach Rates

Isotope	$g/cm^2/day$
Cs-137	$3\text{-}15 \times 10^{-8}$
Sr-90, Ru-105	$3\text{-}10 \times 10^{-6}$
Co-60	$1\text{-}20 \times 10^{-6}$
Pu-238, -239, Am-241	$1\text{-}5 \times 10^{-8}$

and the size of the salt particles. In other words, products with coarser crystals or higher concentrations are leached more easily. Because the extruder-evaporator leads to extremely homogeneous end products with average particle sizes between 10 and 30 μ, asphalt-salt products are 100-200 times less leachable than products from other commonly used processes.

CONCLUSIONS

In summation, the WPC-VRS system, using plastic binders, provides the producers of hazardous wastes with state-of-the-art technology in a system of proven reliability. When this system is selected and incorporated in an actual processing plant, the operator will have at his hand all advantages of the WPC-VRS system. These advantages are there for his use, and it is expected that all or most of the benefits described below will be realized at each installation.

One-Step Volume-Reduction and Solidification

The WPC-VRS system will provide a significant reduction in the volume of waste that requires disposal as compared to other nonvolume-reduction solidification systems. Considering the mixture of wastes produced by industry, an overall volume-reduction factor of 5-20 will be achieved with the WPC-VRS system when compared to cement systems which are usually volume-increasing.

The WPC-VRS system uses an extruder-evaporator to remove the free liquid from waste materials and simultaneously disperse the remaining waste solids as microscopic particles into an asphalt matrix, which naturally solidifies in its container on cooling.

Waste Disposal Costs

Disposal and handling costs are significantly reduced when the WPC-VRS system is used. The initial capital and operating costs are reduced for containers and container storage area, because fewer containers are required. The cost of solidification agent, transportation costs and disposal fees will also be drastically reduced, because of the reduced waste volume requiring disposal.

Reliability

Reliability of the extruder-evaporator has been proven in many applications to be extraordinarily high. The machine is routinely used for plastics compounding applications in which the required availability of this equipment must be 98% or greater. Similar reliability has been seen in hazardous waste processing applications where the machines have operated without major servicing for over 15,000 hours of operation.

The development of the WPC-VRS system is a result of many years of extensive R&D in Europe and the U.S., coupled with conservative design and long operating experience. The process is primarily a mechanical one and its simplicity of operation further enhances reliability.

Variety of Inputs

A wide variety of concentrations of chemicals, slurries and other residues from industry are all capable of being handled by the WPC-VRS system. At the option of the operator, waste liquids and solids can be fed to the unit either individually or simultaneously. The constant kneading and mixing of the extruder-evaporator assures a homogeneous, monolithic end product that does not stratify, settle or produce pockets of concentrated contaminants. In addition, the end product contains no free liquid to increase the volume or promote leaching.

Operation

The system can be totally automated or any degree of manual operation can be provided, depending on client requirements. The extruder is designed for full-time, routine operation. It is versatile in that on/off operation can also be a routine procedure. The system does not require continuous monitoring by operators. The system requires operator attention only during start-up and shutdown, and when containers are moved between the stations of the solidification and storage area. Thus, the system operates on a schedule comparable to and compatible with the operation of other waste treatment equipment.

Safety

The extruder-evaporator has a long history of safe operation in the plastics and chemical industries with over 2000 units in operation worldwide for decades. It is a dust-free system, and its operation at near ambient pressure virtually eliminates the potential for in-process accidents, leaks and spills. Failsafe automatic controls are provided. The system also operates at a temperature nearly 300°F (150°C) below the ignition point of asphalt, thereby eliminating the probability of fire. Automatic annunciators and safety interlocks for all critical, safety or operating parameters are standard features.

Maintenance

Extremely low maintenance costs result because of fewer system components. At Karlsruhe, West Germany, the system operates routinely on an 8-hr/day, 5-day/week schedule for 50 weeks every year, with all maintenance performed during a scheduled two-week shutdown. The system is designed to operate continuously for at least one year without maintenance. Since 1965 at Marcoule, France, CEA has successfully operated two systems processing over 20,000 drums in more than 25,000 hours of operation with no mechanical failure. In plastics compounding applications, the WPC extruder has operated continuously for thousands of hours without a shutdown for maintenance.

Use of Standard Containers

Standard 55-gal drums, 50-ft^3 liners or other specialty forms can be used. Therefore, if future disposal methods should require a change in container size or configuration, the same system could be used with the new containers with only minor equipment and interface modifications required.

Solidification Agent

The physical, chemical and environmental properties of asphalt are well known and documented as a solidification agent for radioactive wastes. Asphalt is acceptable for burial at all major U.S. and European burial sites. Asphalt is resistant to both bacterial action and leaching, and therefore is perhaps the most environmentally acceptable immobilizing agent available. In addition, asphalt is a petroleum by-product that is available with a stable, low cost, and is compatible with most wastes. As with all thermoplastic materials, asphalt requires no chemical reaction for solidification.

Ultimate Liability

Due to the known reliability of the system to produce a homogeneous end-product that is resistant to environmental conditions, the originator/producer of the waste is able to use BAT for their final waste treatment prior to ultimate disposal. This treatment provides assurance that the long-term effects and/or liability will be minimized.

Onsite Storage

If it is required that solidified wastes must be stored onsite, asphalt proves to be far superior to other solidification agents. The volume reduction capabilities of the WPC-VRS system provide approximately an 80% savings in the amount of onsite storage space required.

REFERENCES

1. Doyle, R. D. "Use of an Extruder/Evaporator to Stabilize and Solidify Hazardous Wastes," in *Toxic and Hazardous Waste Disposal, Vol. 1*, R. B. Pojasek, Ed. (Ann Arbor, MI: Ann Arbor Science Publishers, Inc., 1979), pp. 65–92.
2. Johnson, J. C., and R. L. Lancione. "Laboratory Assessment of Fixation and Encapsulation Processes for Arsenic-Laden Wastes," presented at Hazardous Waste Land Disposal Symposium in San Antonio, TX, March 8, 1978.
3. "Bituminization of Radioactive Wastes," Technical Reports Series No. 116, International Atomic Energy Agency, Vienna, STI/DOC/116 (1970).
4. Doyle, R. D. and J. E. Stewart. "Volume Reduction and Solidification Using Asphalt," presented at Waste Management '78, Tucson, Arizona, March 8, 1978.

CHAPTER 12

ULTIMATE HAZARDOUS WASTE DISPOSAL BY INCINERATION

John C. Reed and Berkley L. Moore
Illinois Environmental Protection Agency
Springfield, Illinois

INTRODUCTION

Incineration is a very attractive option in implementing hazardous waste disposal in the U.S. Many state and federal regulations use the approach of measuring incineration efficiency by the concentration of carbon monoxide (CO) in the incinerator's stack gases. However, available data from several incineration processes shows no correlation with destruction of hydrocarbons. This could be inferred from basic kinetic considerations since the kinetics of formation of CO are quite different from the kinetics of destruction of hydrocarbons. The available data on hydrocarbon destruction can be approximated by a first-order kinetic expression. The U.S. Environmental Protection Agency (EPA) proposal on incineration of hazardous wastes is discussed in relation to this approach and additional suggestions are made for checks on incinerator operation and permissible ground-level concentration.

GENERAL CONSIDERATIONS

In general, a hazardous waste can be defined as any waste with the potential for acute or chronic adverse health or ecological effects when its disposal is uncontrolled. EPA has published a proposal which separates such wastes into four categories: ignitable, corrosive, reactive and toxic [1]. The main concern here, in disposal by incineration, is with toxic wastes, especially those containing chlorinated hydrocarbons. Such wastes are not hazardous if properly treated but they must be evaluated carefully and a proper disposal method must be used. Only about 5% of the materials in the refuse waste stream are characterized as hazardous [2]. Those wastes that cannot be incinerated, such as heavy metals, can be placed in a secure landfill. Safeguards are built into landfill leachates containing heavy metals due to cation exchange and precipitation in soils. However, chemical and microbial interactions are often insufficient to degrade synthetic organic chemicals (particularly polycyclic or chlorinated hydrocarbons). Although a number of materials have been studied for subsurface barriers (e.g., natural subsoil and manmade materials), none can seal for an indefinite period of time. Once a barrier has failed, the failure is difficult to detect due to small horizontal spread in the groundwater.

Although it can be expensive, incineration may be the most environmentally sound option for implementing hazardous waste disposal in the U.S., providing the problems associated with incineration are assessed and dealt with on an individual basis. There is basically no limit to the type of organic material that can be disposed of by incineration provided sufficient auxiliary fuel is used. To reduce fuel costs, the typical practice is to burn highly combustible material (such as used solvents) with low- or noncombustible materials (such as aqueous solutions containing hazardous organic materials). Unfortunately, this practice can require temporary storage of hazardous materials which then leads to leakage, spillage or other unsafe conditions and ultimately causes more hazard than the incineration process itself. Another problem is the high temperatures frequently required to ensure complete destruction of hazardous materials. These temperatures require special construction materials, special precautions to avoid thermal shock and special inspection and maintenance procedures.

The EPA proposal [1] has addressed these concerns and has suggested standards for storage (Section 250.44), inspection (Section 250.43-6) and incineration (Section 250.45-1). A general discussion of incineration and field inspection procedures was published earlier by the EPA [3].

HAZARDOUS WASTE INCINERATOR EMISSIONS

There have been a number of tests and research and development studies on destruction of hazardous materials which were recently summarized in an EPA publication [4]. From a quantitative incinerator operation standpoint, the most useful references are those describing the incinerator ship Vulcanus [5, 6], a Midwest Research Institute (MRI) experimental incinerator [7] and several large-scale commercial incinerators [8-10]. The Vulcanus burned a mixture of chlorinated C_3 to C_6 nonaromatic hydrocarbons, the MRI incinerator burned specific pesticide formulations and the commercial incinerators burned mixed chemical process wastes (ethylene waste, C_5-C_6 waste, phenol waste, methyl methacrylate waste, polyvinyl chloride (PVC) waste, polychlorinated biphenyl (PCB) capacitors and nitrochlorobenze waste). Although the test information is incomplete, certain generalizations can be postulated from the data.

First, contrary to a common current assumption, the CO concentration in the effluent gas does not correlate with the degree of destruction of hydrocarbons. Figure 1 shows that high values of CO do not correspond with high values of effluent/charge hydrocarbon ratios. In general, a submitted evaluation of the data shows no correlation between effluent/charge hydrocarbon ratio and outlet CO concentration. This observation seems reasonable from a

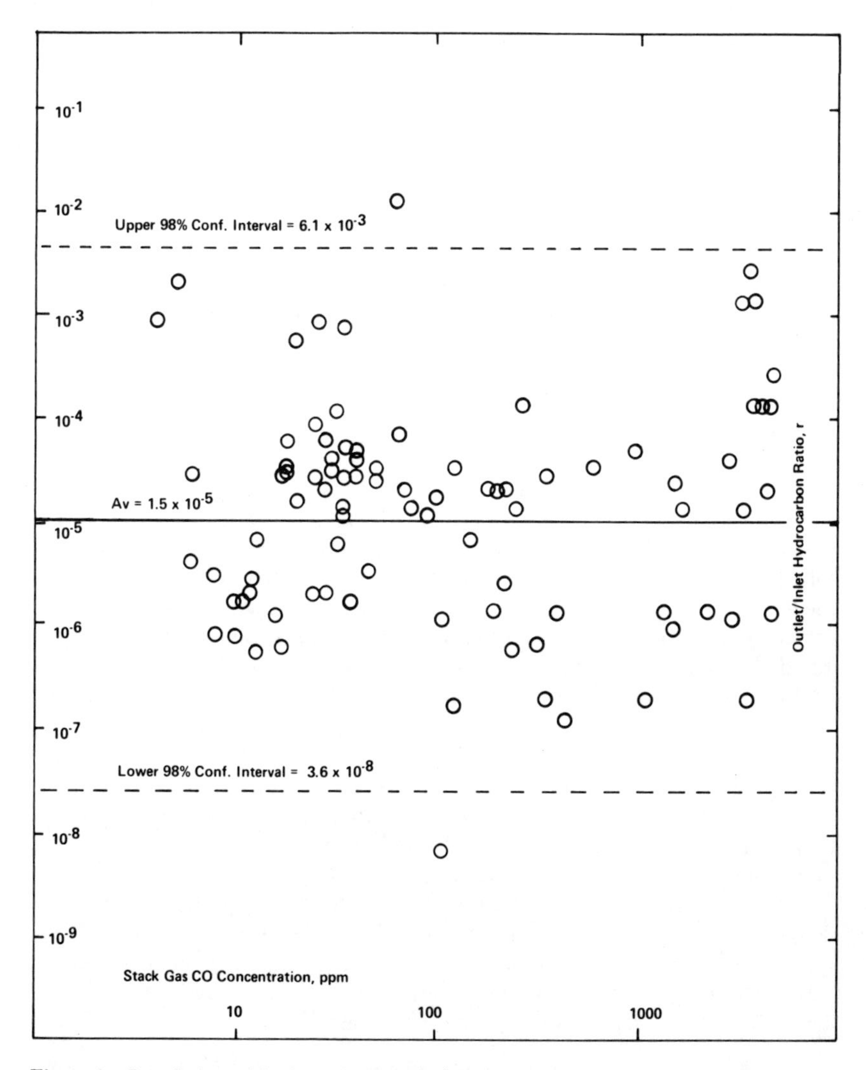

Figure 1. Correlation of hydrocarbon outlet/inlet concentration ratio with carbon monoxide concentration.

kinetics viewpoint since it would not be expected that the kinetics of formation of CO would be the same as of the destruction of the original compound.

However, a first-order kinetic rate expression can be used to evaluate the destruction of organic compounds. The complex reaction sequence in the incinerator can be simplified to the first-order kinetic expression:

$$dC/dt = -kE \qquad (1)$$

where C = concentration of waste organic oxidized
k = kinetic rate constant
t = time, sec

The solution to this equation can be given as:

$$ln\,r = -kt \qquad (2)$$

where r = concentration ratio of effluent organic material/concentration of entering organic material
t = incinerator residence time, sec

The Arrhenius expression can then be used to express the temperature dependence of the reaction rate constants of Equation 2 for each concentration ratio-residence time for each incinerator:

$$k = A\,exp(-E/RT) \qquad (3)$$

where A = kinetic rate constant intercept
E = activation energy, cal/mol
T = absolute temperature, °K

Figure 2 shows a correlation of the kinetic rate constants as a function of temperature for all the available incinerator performance data [4-10]. The solid line in Figure 2 represents the least-squares correlation of the data and the dotted lines indicate the 95% upper and lower confidence level of the correlation. The least-squares correlation for the activation energies (E) and kinetic constant intercepts (A) for Equation 3 are given in Table I.

The least-squares equation for all the incinerators was:

$$k = 86\,exp(-8580/RT) \qquad (4)$$

which had a correlation coefficient of 0.8. All of the activation energies and kinetic intercepts reported were much lower than values obtained in laboratory experiments for other compounds [11]. This is probably due to the more complex flow and kinetic interactions occurring.

However, Equation 4, which represents a wide variety of incinerator designs and operating conditions, can be a useful empirical tool for determining the suitability a given temperature residence time combination. Since this equation represents an average kinetic constant, the 95% lower confidence limit can be represented approximately as:

$$k = 27.1\,exp(-8580/RT) \qquad (5)$$

which may be more appropriate for design analysis.

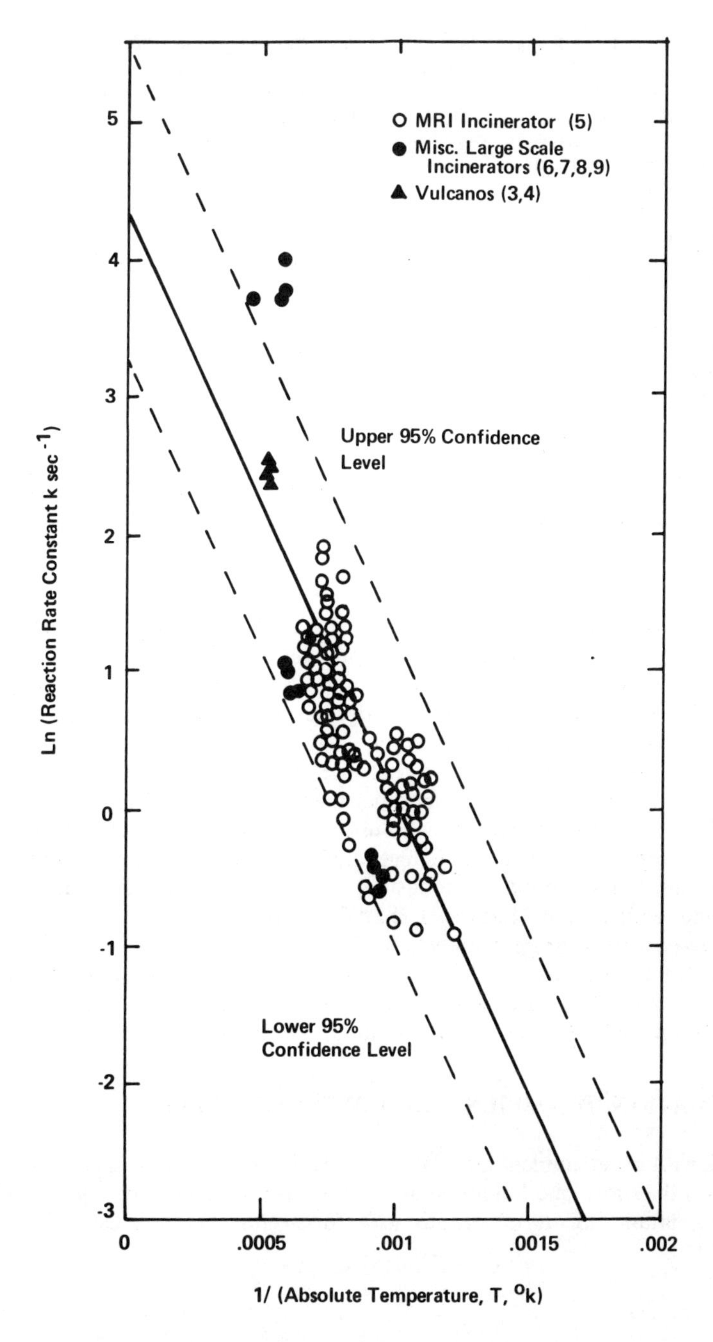

Figure 2. Correlation of reaction rate constant with absolute temperature.

Table I. Conversion of Hazardous Wastes by Incineration

Material	Activation Energy (cal/mole)	Kinetic Intercept A
Chlorinated Hydrocarbon	3,970	35.30
10–25% DDT	294	3.44
19–41% Aldrin	7,720	57.60
10–22% Picloram	10,400	158.00
25–57% Malathion	3,510	7.05
20–60% Toxaphene	3,540	8.40
41–80% Atrazine	5,010	17.80
50% Captan	6,720	44.20
75% Zineb	7,330	56.90
0.3% Mirex	7,200	31.10
Misc. (PCB Capacitors, Nitrochlorinated Waste, Ethylene Waste, Chlorinated Pentadiene)	15,700	1,461.00
Overall	8,580	86.00

Considering the recommendation of EPA for a combustion temperature of 1000°C at 2 sec retention time [1], combining Equations 2 and 4 would give $r = 0.003$ which would be an average combustion efficiency for most wastes of 99.7%. In the case of halogenated organics, the EPA recommendation of 1200°C and 2 sec residence time would give, on combining Equations 2 and 5, $r = 0.05$. (Note that Equation 5 rather than Equation 4 was combined with Equation 2 to obtain a minimum rather than an average efficiency.)

This would give a minimum combustion efficiency of 95%, which would not be acceptable with certain halogenated organics, especially pesticides and herbicides. For a minimum combustion efficiency of 99.9%, Equations 2 and 5 would require a temperature of 1820°C (3310°F).

However, the average combustion efficiency using Equations 2 and 4 with 1200°C and 2 sec would be 99.99%. Thus it may be possible to operate an incinerator at less than 1820°C and 2 sec residence time, but it would be prudent to establish effluent checks for unburned organics.

ESTIMATION OF AMBIENT AIR CONCENTRATIONS

As with other ambient air quality standards, there must be a consideration of both the short- and long-term air quality. As pointed out by Turner [12], the maximum short-term concentration for a Gaussian plume can be given as:

$$G = (0.117Q/U\ S_Y\ S_Z) \times 10^6 \tag{6}$$

which occurs at

$$S_Z = \sqrt{2}H$$

where G = maximum ground level concentration, $\mu g/m^3$
Q = source emission strength, g/sec
S_y = dispersion coefficient in horizontal direction, m
S_z = dispersion coefficient in vertical direction, m
U = wind speed, m/sec
H = effective stack height, m

Inspection of Turner's Figure 4-1 indicates that the product of the dispersion coefficients $S_y \cdot S_z$ can be approximated for distances of a kilometer or less as:

$$S_y S_z = 500\ D^{1.74}\ \text{(F Stability)} \tag{7}$$

$$= 50{,}000\ D^{2.15}\ \text{(A Stability)} \tag{8}$$

and Turner's Figure 3-3 indicates

$$S_z = 14.7D^{0.814}\ \text{(F Stability)} \tag{9}$$

$$= 192D^{1.15}\ \text{(A Stability)} \tag{10}$$

where D = distance from source (km). Combining Equations 7 through 10 with Equation 6, and assuming a minimum wind speed of 2 m/sec gives:

$$G = 17{,}500QH^{-2.14}\ \text{to}\ 11{,}400QH^{-1.87} \tag{11}$$

The long-term air quality can be examined by the screening parameter:

$$P = GD/Q^{2/3} \tag{12}$$

where P = dispersion parameter for climatological dispersion modeling results
G = ground level concentration, mg/m^3
Q = source emission strength, g/sec
D = distance from source, km

It was found that this parameter was a normally distributed function with a median value of 1.48 and standard deviation of 0.52 [13]. By the use of this parameter it is possible to predict the contribution of a given source with any designated degree of probability. For example, using the most likely value of the parameter, Equation 12 can be used to provide the most probable contribution of a source:

$$G = 1.48Q^{2/3}/D \tag{13}$$

In the case of hazardous materials, it may be necessary to set conditions so that the probability of an environmental impact is very small.

For example, the only chance in a hundred that P will be greater than 2.66 so that Equation 12 can be given as:

$$G = 2.66Q^{2/3}/D \tag{14}$$

In the absence of a given standard the value of G is often taken as a fraction of the allowable workplace values such as those given by the Occupational Safety and Health Administration (OSHA) or the TLV values of the

American Conference for Governmental Industrial Hygienists (ACGIH) (1/30 to 1/100 have been fractions used in the past). For example, if the OSHA continuous exposure level for compound X is 500 $\mu g/m^3$, the permissible long-term average emission rate with the nearest receptors at 1 kilometer would be:

$$Q = \left[\frac{500}{(2.66)(100)}\right]^{3/2} = 2.6 \text{ g/sec}$$

This emission rate could then be related back to the incinerator feed rate by Equation 2 and 4 for the average incinerator operating temperature and residence time. Considering 1000°C and 2-sec residence time gives an incinerator feed rate of 2.6/0.003 = 867 g/sec of undiluted organic compound. If the feed contains other compounds which have exposure limits, the EPA recommendation is that the exposure to the sum of these compounds must not exceed unity as given by [15]:

$$C_i/L_i = 1 \tag{13}$$

where C_i = concentration of component i
L_i = exposure limit of component i

It should be emphasized that these equations are a screening device to discard definitely unsatisfactory possibilities, but modeling may be needed for those cases that are borderline.

INCINERATOR OPERATING PROCEDURE SPECIFICATION

The record-keeping system proposed by EPA [1] requires the facility to keep an operating log which potentially could include the chemical composition, ultimate analysis, heat content, burning rate, combustion temperature and chemical composition of the exhaust gases. The requirements specifically refer to "... a detailed chemical and physical analysis (to) identify the hazardous characteristics of the waste which must be known to comply with the requirements of this subpart or with the conditions of a permit ..." [1]. Unfortunately, this statement is vague as to exact requirements and must be amplified by a set of explicit permit conditions. Being too vague in specifying operating procedures leads to unnecessary misunderstanding as to requirements as well as unfavorable public reaction if a health hazard incident occurs. It is quite reasonable that the general requirements should be flexible; however, each incinerator must be considered quite carefully on an individual basis.

Appendix A gives a list of possible permit specifications of operating conditions that have evolved through a series of conditions for different hazardous waste incinerators. The basic philosophy is to obtain sufficient information to determine significant deviations from normal operation. The blank items on conditions c, d, h and i would be completed by general considerations such as given in the preceeding sections with possible modification based upon the test burn(s). The requirements in general parallel those of the EPA proposal with a scrubber log substituted for CO testing and monitoring, since CO concentration does not seem to be directly related to incinerator performance. The presence of detectable amounts of unburned organochlorine compounds or other pesticides, etc., would be an indication of improper incineration. In addition, any buildup of dissolved solids, suspended solids, heavy metals or a pH drop would be an indication of improper scrubbing.

Although it may not seem justified to place such emphasis on checking scrubber performance, a recent series of tests on the scrubbing of phosgene emissions gave actual efficiency at least one order of magnitude below design level [13]. This result was attributed to depletion of the caustic by a reaction with both phosgene and atmospheric carbon dioxide.

REFERENCES

1. "Hazardous Waste Proposed Guidelines and Regulations and Proposal on Identification and Listing," *Federal Register, 43*:58947 (1978).
2. Barnhart, B. J. "The Disposal of Hazardous Wastes," *Environ. Sci. Technol.*, 12:1132–1135 (1978).
3. Devitt, T. W., R. W. Gerstle and N. J. Kulugian. "Field Surveillance and Enforcement Guide: Combustion and Incineration Sources," APTD-1449, U.S. EPA, Office of Air Quality Planning and Standards, Research Triangle Park, NC (1973).
4. Wilkinson, R. R., G. L. Kelso and F. C. Hopkins. "State-of-the-Art Report–Pesticide Disposal Research," U.S. EPA, Municipal Environmental Research Laboratory, Cincinnati, OH EPA-600/2-78-183 (1978).
5. Wastler, T. A., C. K. Ofutt, C. K. Fitzsimmons and P. E. Des Rosiers. "Disposal of Organchlorine Wastes by Incineration at Sea," U.S. EPA Office of Water and Hazardous Materials, Washington, DC, EPA-430/9-75-014 (1975).
6. Clausen, J. F., H. J. Fisher, R. J. Johnson, E. L. Moon, C. C. Shih, R. F. Tobias and C. A. Zee. "At-Sea Incineration or Organchlorine Wastes Onboard the M/T Vulcanus," U.S. EPA, Office of Energy, Minerals and Industry, Research Triangle Park, NC EPA-600/2-79-196, (1977).
7. Ferguson, T. L., F. J. Bergman, G. R. Cooper, R. T. Li and F. I. Honea. "Determination of Incinerator Operating Conditions Necessary for Safe Disposal of Pesticides," Contract No. 68-03-0286, National Environmental Research Center, Cincinnati, OH (1975).
8. Clausen, J. F., R. J. Johnson and C. A. Zee. "Destroying Chemical Wastes in Commercial Scale Incinerators. Facility Report No. 1–The Marquardt Company," PB-265541, U.S. Dept. of Commerce, NTIS (1977).
9. Ackerman, D. et al. "Destroying Chemical Wastes in Commercial Scale Incinerators. Facility Report No. 6–Rollins . . . ," PB-270897, U.S. Dept. of Commerce, NTIS (1977).
10. Ackerman, D. G. et al. "Destroying Chemical Wastes in Commercial Scale Incinerators. Final Report–Phase II," Contract No. 68-01-2966, Arthur D. Little Co., Cambridge, MA (1977).

11. Lee, K., H. J. Jahnes and D. C. Mccauley. "Thermal Oxidation Kinetics of Selected Organic Compounds," Air Pollution Control Association 71st Annual Meeting Paper 78-58.6 (June 25–30, 1978).
12. Turner, D. B. "Workbook of Atmospheric Dispersions Estimates," U.S. EPA Publication No. AP-26, Research Triangle Park, NC (1970).
13. Kistner, S., D. Lillian, J. Ursillo, N. Smith, K. Sexton, M. Tuggle, G. Espisota, G. Podolak and S. Mallen. "A Caustic Scrubber System for the Control of Phosgene Emissions: Design, Testing and Performance," *Air Poll. Control Assoc. J.* 28:673 (1978).

APPENDIX A

EXAMPLE PERMIT CONDITIONS FOR GENERAL HAZARDOUS WASTE INCINERATOR

A. The permittee will maintain, on each incinerator individually, a permanent daily log of the materials burned. These logs shall include the following information:
   1. Sufficiently detailed analysis of the waste stream to indicate any change in waste characteristics, including but not limited to physical appearance, specific gravity, pH and vapor pressure. If a change is noted in the waste characteristics, a complete analysis shall be made, including chemical composition and ultimate analysis with specific determinations of halogen, sulfur, heavy metal, beryllium, asbestos and ash contents.
   2. Rates of materials burned during each day of operation expressed in pounds of material burned per hour.
   3. Heat content of materials burned during each day of operation expressed in Btu/lb of material burned.
   4. Orsat or equivalent analysis of exhaust gases during each day of operation.
   5. Total hours of operation during each day of operation.
   6. Type of any auxiliary fuels used for firing, and purposes and sulfur content of any such fuels (expressed in percentage by weight by fuel used).
   7. Firing rate of any auxiliary fuels used expressed in pounds of fuel burned per hour.
   8. Total amount of each material burned each day.
   9. Operating parameters of the scrubber system, including but not limited to:
      a. chemical composition of inlet and outlet scrubbant, including but not limited to chloride, residual chlorine, sulfate, total sulfide, calcium, amines, organchlorine pesticides and herbicides, and mercaptans.
      b. pH and total alkalinity of inlet and outlet scrubbant.
      c. suspended solids, heavy metals and beryllium content of inlet and outlet scrubbant.
      d. total dissolved solids (TDS) of inlet and outlet scrubbant.
      e. flowrate of scrubbant.
      f. inlet and outlet temperature of gas to scrubber system.
      g. pressure drop across the packed bed and deminster portion of the scrubber system.

   At any time during hours of operation or normal working hours, any representative of the agency shall have the right and authority to inspect and photocopy the daily logs required to be maintained under this permit condition.

B. The permittee shall submit a report to the agency once every 90 days with the first such report due 90 days after the date of this permit. Each such report shall contain a summary of the information required to be maintained under permit condition A.

C. The permittee shall burn only wastes as defined in ____. Such wastes shall contain less than ____% (by weight) chlorine, less than ____ ppm of heavy metals (e.g., lead, cadmium, etc.) at any time and less than ____ ppm beryllium at any time. The burning rate of the ____ wastes shall not exceed ____ lb/hr.

D. A minimum afterburner chamber temperature of ____°F will be achieved prior to the injection of any materials to be burned and such temperature shall be maintained at all times during the burning of the materials. For those materials listed under special condition(s) below, the afterburner chamber temperature shall be maintained at the level specified in the written request for approval.

E. Prior written approval from the agency shall be required for the burning of materials containing any of the following:
   1. Toxic compounds of phosphorus or nitrogen and all mercaptans.
   2. Pesticides and herbidices.
   3. Any material for which a final rule has been promulgated by the EPA Administrator under the Toxic Substances Control Act (TSCA) or the Resource Conservation and Recovery Act (RCRA).
   4. Any material(s) whose individual or total compositional change in hazardous components (as defined by the EPA Administrator under TSCA or RCRA) is more than 10% from the original test burn composition as given in the originally submitted test data.

F. When requesting approval, the permittee shall provide, as a minimum, the following information:
   1. A description of the material, including but not limited to toxicity, chemical structure, manufacturer and trade name.
   2. A chemical analysis of the material, as defined in condition A.1.
   3. A description of the disposal method for the scrubbant and all solid wastes generated. If the solid waste is to be disposed of at a landfill, the permittee shall complete Form ____.
   4. Afterburner chamber temperature to be achieved and maintained.
   5. Incineration rate of material to be burned in lb/hr and total amount of material(s) to be burned.
   6. Projected operating parameters of the scrubber system in accordance with the requirements of condition A.9.
   7. Firing rate and type of auxiliary fuel as described in conditions A.6 and A.7.
   8. Projected operating parameters of the scrubber system in accordance with requirements of condition A.9.
   9. Test results of the trial burn giving the information required under condition A.

G. Before commencement of operation, test burn(s) shall be made of the waste(s) to be incinerated and the information required under condition A shall be forwarded to the agency immediately after the test results are compiled and finalized.

H. The afterburner chamber temperatures shall be monitored continuously by an instrument which produces a continuous permanent readout of such monitoring. This instrument shall be capable of full-scale response time of less than 1 sec and be capable of measuring ____°F within an accuracy of ±10°F. The instrument shall be calibrated at a minimum frequency of once every 15 days. A permanent record shall be maintained of each calibration. All original monitoring records including all recording or readout charts and calibration records shall be maintained for inspection and photocopying by the agency.

I. The scrubbing liquid shall contain less than ____ mg/l of hydrocarbon.

CHAPTER 13

OCEAN INCINERATION OF TOXIC CHEMICAL WASTES

Max Halebsky
Global Marine Development Inc.
Newport Beach, California

INTRODUCTION

A study was performed on the economic and environmental viability of incinerating toxic chemical wastes at sea using a U.S. flagship(s). Such incineration has been conducted in Europe since 1969. There have also been several burns of U.S. wastes in the Gulf of Mexico and the Pacific Ocean from 1974 to 1977 using the European incineration ship Vulcanus. The impact of international, national and state regulations on waste quantities and ocean incineration was evaluated. The United Nations' Inter-governmental Maritime Consultative Organization (IMCO) established international requirements for ocean incineration in October 1978 and is now preparing incineration technical guidelines as well as ship design requirements. Toxic waste quantities available for ocean incineration (primarily chlorinated and unchlorinated organics) were estimated for 1977, 1983 and 1989 for the United States, Canada, Mexico, Brazil, Japan and Western Europe. Capital cost, operating cost and incineration cost per ton of waste were calculated for a U.S. flagship and compared with incineration charges for a European ship and for land-based incineration. The conclusion is that a U.S. incineration flagship(s) is economically and environmentally viable.

REGULATIONS

There are existing and planned international conventions, federal laws and regulations, and state laws and regulations which directly and indirectly pertain to incineration ships and influence the viability of ocean incineration.

The legal and regulatory requirements for incineration ships can be considered in two categories—those pertaining to the incineration process such as incinerator performance and the exhaust from the incinerator, and those pertaining to the ship itself including its design, construction and operation. In addition, federal and state requirements on waste generation, discharge, transportation, storage and disposal will have an impact on both the quantity and type of waste reasonably available for ocean incineration and the associated costs for transferring the wastes from the originator's plant to the incineration ship.

International Conventions

Incineration Process

On October 12, 1978, IMCO adopted "Regulations for the Control of Incineration of Wastes and Other Matter at Sea" as an addendum to the "Convention on the Prevention of Marine Pollution by Dumping of Wastes and Other Matter" [1]. Action leading to the regulations was initiated with an IMCO meeting in London in March 1977 at which the U.S. presented technical incineration guidelines for consideration. The guidelines were later divided into mandatory and guideline sections and the mandatory section was adopted as the IMCO regulations in London on October 12, 1978. The guideline section was planned for adoption in 1979. The regulations and guidelines only address liquid organic wastes with emphasis on chlorinated organics since there have been no successful burns of other types of toxic wastes or of wastes in solid form at sea.

Highlights from the adopted regulations are contained in Table I. In addition to those requirements, the guidelines call for a 1-sec residence time at 1250°C and a minimum of 3% excess oxygen in the stack exhaust.

Incineration Ship Design and Construction

There is no international convention on the design and construction of incineration ships. However, there is an IMCO convention, the Bulk Chemical Code [2], which deals with transporting dangerous chemicals in bulk, including chlorinated organics, which are prime candidates for incineration at sea. Accordingly, the IMCO incineration guidelines state that incineration vessels should carry a valid "Certificate of Fitness" under the Bulk Chemical Code and should comply with the requirements for a Type II ship as defined in that code.

Federal Regulations

Incineration Process

There are no U.S. laws or regulations specifically directed at incineration at sea. However, the U.S. Environmental Protection Agency (EPA) issued special permits to the Shell Chemical Company for incineration of chlorinated organic liquids in the Gulf of Mexico using the Vulcanus and designated a

Table I. Incineration Regulation Highlights

- "Incineration at sea" means the deliberate combustion of wastes or other matter on marine incineration facilities for the purpose of their thermal destruction.
- Combustion of wastes shall be at a minimum of 1250°C.
- There shall be no black smoke nor flame extension above the plane of the stack.
- Combustion efficiency shall be routinely measured on incinerator vessels and shall be at least 99.95 ± 0.05%. Incinerator design, as determined during the initial vessel survey, must demonstrate in excess of 99.9% combustion and hydrocarbon destruction efficiency.
- As part of the initial survey of a proposed marine incineration system, ensure that the feed of waste to the incinerator is automatically shut off if the temperature drops below approved minimums.
- The incineration system shall be surveyed at least every two years to ensure that the incinerator continues to comply with the regulations.

specific ocean site in the Gulf of Mexico for such incineration [3]. The IMCO requirements and essentially all of the guidelines will be converted by the EPA into regulations which will be published in the *Federal Register* and then implemented. These regulations will be issued under PL 92-532, the Marine Protection, Research and Sanctuaries Act (MRPSA) of 1972.

Incineration Ship Design and Construction

Additional and/or revised regulations that will affect incineration ship construction and certification will result from the International Conference on Tanker Safety and Pollution Prevention [4], which was held on February 9-16, 1978, under the auspices of IMCO. Areas involved include segregated ballast, inert gas systems, steering systems and radar. The Coast Guard's implementation intent was published in the April 20, 1978, issue of the *Federal Register* [5].

Chapter I of Title 46 of the Code of Federal Regulations, and especially Subchapter "O" entitled "Certain Bulk Dangerous Cargoes" consisting of Parts 151-154, contain the U.S. regulations which correspond to the IMCO Bulk Chemical Code. Of special importance is Part 153–Safety Rules for Self-Propelled Vessels Carrying Hazardous Liquids. These regulations were updated as of the September 26, 1977, issue of the *Federal Register* [6] and implement the IMCO code.

Related Environmental Activities

Resource Conservation and Recovery Act (RCRA) [7]. Potentially the strongest impetus to ocean incineration is RCRA. The implementing regulations will govern land waste disposal from generation, to transportation, to storage, and disposal, and will establish stricter requirements than presently exist. This will make land disposal more expensive and less attractive. The last of the proposed regulations was published in the December 18, 1978, issue of the *Federal Register.* The law takes effect six months after the final regulations are promulgated. The best current estimate is that the effective date will be July 1980 barring any delaying court challenges.

The serious intent of EPA to implement RCRA can be seen in the issuance on December 16, 1977, of the draft document, "Strategy for the Implementation of the Resource Conservation and Recovery Act of 1976" [8]. The anticipated result is more effective waste management and an improved potential for ocean incineration due to increases in waste quantities, significant increases in land disposal costs and strong opposition to land disposal sites.

Title 33 CFR Parts 126, 154 and 156. The U.S. Coast Guard (USCG) announced its intent on April 10, 1978 [9], to establish new environmental and safety requirements for waterfront facilities including requirements for handling, storage, loading and unloading of hazardous materials. This would include toxic chemical waste and could have a cost impact on the waterfront aspects of incineration ship operations. The new regulations will represent an updating, consolidation and detailed expansion of the existing regulations in 33 CFR 126, 154 and 156, and will be in compliance with the Magnuson Act (50 USC 191), the Ports and Waterways Safety Act (86 Stat. 424, 33 USC 1221-7), the Clean Water Act and the Transportation Safety Act, which includes the Hazardous Materials Safety Act, (88 Stat. 2156, 49 USC 1801 et seq.).

WASTE STREAMS AND QUANTITIES

Four industries were selected for determination of waste streams and potential ocean incineration quantities based on prior EPA studies. These prior studies divided industry into 15 groups as shown in Table II and assessed the waste practices in those 15 groups. The types and quantities of hazardous wastes were established as of 1973, including the technology and costs for disposal of the wastes. The assessment studies showed that four of the industry groups produced significantly greater quantities of more hazardous waste than the others. As a result, an additional round of studies was funded by the EPA for those four groups. Alternative waste treatment methods and costs were investigated in greater depth than in the prior studies and the costs were expressed in 1976 dollars. The basis for the alternative methods was another EPA study [10], the purpose of which was to identify environmentally acceptable alternatives to current disposal methods. This study confined itself to:

1. organic chemicals and pesticides;
2. inorganic chemicals;
3. metal smelting and refining; and
4. petroleum refining.

Table II. Industrial Waste Assessment Groups

Organic Chemicals, Pesticides and Explosives Industries
Paint and Allied Products Industry, Contract Solvent Reclaiming Operations and Factory Application of Coatings
Treatment and Disposal in the Pharmaceutical Industry
Battery Industry
Electronic Components Industry
Electroplating Industry
Petroleum Refining Industry
Rubber and Plastics Industry
Primary Metals (Mining and Processing) Industry
Secondary Refining and Smelting Industry
Special Machinery Industry
Leather Tanning Industry
Textile Industry
Oil Re-refining Industry
Inorganic Chemicals Industry

Organic Chemicals and Pesticides

Representative organic chemical (Standard Industrial Classification (SIC) categories 2861, 2865 and 2869) and pesticide (SIC categories 2879 and 28694) hazardous waste streams are shown in Table III and are based on a previous EPA assessment study [11]. These waste streams were examined in a follow-up EPA alternatives study [12] which determined which treatment processes would represent improvements and could be used as alternative waste disposal methods.

The alternative methods are environmentally acceptable and meet one or more of the following three criteria:

- resource and/or energy recovery
- detoxification
- volume reduction

The results of the study [12] are summarized in Table IV and include the following items:

- Waste stream components for 16 organic chemical waste streams and 5 pesticide waste streams including yearly waste quantities generated for the indicated typical plant sizes.
- Alternative treatment processes and degree of development of these processes as shown by the Roman numeral category. Table V defines the categories.

Table III. Representative Hazardous Waste Streams–Organic Chemicals and Pesticides Industries [10]

Waste Stream No.	Description
	Organic Chemicals Industry
1	Heavy ends from purification columns, perchloroethylene manufacture
2	Heavy ends from purification column, nitrobenzene manufacture
3	Solid tails from solvent recovery system, chloromethane solvent manufacture
4	Heavy ends from fractionator, epichlorohydrin manufacture
5	Centrifuge residue sludge, toluene diisocyanate manufacture
6	Heavy ends from ethylene dichloride recovery still, vinyl chloride monomer manufacture
7	Heavy ends from methanol recovery column, methyl methacrylate monomer manufacture
8	Heavy ends from purification column, acrylonitrile manufacture
9	Still bottoms, maleic anhydride manufacture
10	Lead sludge from settling basin, lead alkyls manufacture
11	Triethanolamine column heavies, ethanolamines manufacture
12	Still bottoms from stripping column furfural manufacture
13	Filter solids, furfural manufacture
14	Spent reactor catalyst, fluorocarbon manufacture
15	Still bottoms from fractionating column, chlorotoluene manufacture
16	Distillation residues from bath fractionating towers, chlorobenzene manufacture
	Pesticides Industry
17	Spent alkali scrubbing solution, cyanuric chloride manufacture in atrazine production
18	Spent activated carbon from adsorption treatment, trifluralin manufacture
19	Filter cake, malathion manufacture
20	Liquid process waste, malathion manufacture
21	Sulfur sludge from chlorination unit, parathion manufacture

- Waste disposal costs based on the alternative techniques, based on acceptable* chemical landfill disposal and acceptable land incineration disposal, including scrubbers.

Inspection of Table IV results in the following observations:

- Alternative treatment costs for organic chemical and pesticide wastes ranged from a return of $378/metric ton of waste due to chemical recovery to a maximum cost of $5560/metric ton for disposal of the wastes. These are fourth-quarter 1976 dollars.
- Cost of chemical landfill disposal for organic chemicals and pesticides ranged from $48 to $326/metric ton, 1976 dollars.
- Cost of land incineration disposal for organic chemicals and pesticides was $30-1240/metric ton, 1976 dollars.
- For the combined total of 21 waste streams, alternative processes were less expensive for 6 streams, chemical landfill was less expensive for 9 streams and land incineration was less expensive for 6 streams. These results and the costs are prior to the impact of RCRA whose implementing regulations will take effect in 1980. RCRA's draft regulations on land disposal impose strict requirements which will mean increased costs for land disposal.

It should be noted that the alternative waste disposal techniques do not, as a rule, eliminate the waste but do reduce the volume so that there is still a finite quantity requiring disposal. This compares with ocean incineration which, as a rule, completely eliminates the waste. Land incineration also usually requires a scrubbing operation which results in an effluent which requires disposition.

Inorganic Chemicals

The inorganic chemical industry produces 19 hazardous waste streams, as identified in a previous EPA assessment study [13]. These are shown in Table VI. Analysis of the data from that study shows that the wastes generally fall into the following four categories:

- sludges from settling ponds, 6-80% moisture
- slurries, 90% moisture
- filter cake, 20-50% moisture
- retort or furnace residues, 10% moisture

The results of the follow-up EPA alternatives study [14] for inorganic chemical wastes are summarized in Table VII. Note that the disposal costs are per metric ton of waste as generated by the manufacturing process. This means that the costs are usually on a wet basis since most of the waste streams contain significant amounts of water. This contrasts to the organic chemical industry where the streams, although often liquids, are considered dry since they are nonaqueous.

*Does not include the potential impact RCRA of 1976 which had not yet been implemented by regulations.

Table IV. Cost Comparison of Alternatives–Organic Chemicals and Pesticides Industries [12] (Basis–Fourth Quarter 1976)

Stream[a] No.	Product and Typical Plant Size (metric ton/yr)	Waste Stream Components	Waste Generation[b] (metric ton/yr)	Alternative Treatment Processes	Alternative Treatment Cost[c] ($/metric ton waste)	Chemical Landfill Cost ($/metric ton waste)	Land Incineration Cost ($/metric ton waste)
			Organic Chemicals Industry				
1	Perchloroethylene 39,000	Hexachlorobutadiene Chlorobenzenes Chloroethanes Chlorobutadiene Tars	12,000	Distillation IV/V	–378[d]	48	45
2	Nitrobenzene 20,000	Crude Nitrated Aromatics	50	Steam Dist., Hydrolysis, Catalysis III/IV	1930	157	NA
3	Chloromethane 50,000	Hexachlorobenzene Hexachlorobutadiene Tars	300	Distillation, Chlorinolysis, Neutralization III/IV/V	646	128	288
4	Epichlorohydrin 75,000	Epichlorohydrin Dichlorohydrin Chloroethers Trichloropropane Tars	4,000	Solvent Extraction, Evaporation, Distillation III/IV	0.50	55	84
5	Toluene Diisocyanate 27,500	Polyurethane Ferric Chloride Isocyanates Tars	558	Hydrolysis, Distillation, Neutralization, Aerated Lagoon III/IV	428	156	231

6	Vinyl Chloride Monomer 136,000	1,2 Dichloroethane 1,1,2 Trichloro-ethane 1,1,1,2 Tetra-chloroethane Tars	1,400	Distillation, Reduction III/IV	–0.86	67	208
7	Methyl Methacrylate 55,000	Hydroquinone Polymeric Residues	4,730	None		76	30
8	Acrylonitrile 80,000	Acrylonitrile Higher Nitriles	160	None		158	350
9	Maleic Anhydride 11,000	Maleic Anhydride Fumaric Acid Chromogenic Compounds Tars	333	None		166	363
10	Lead Alkyls 60,000	Lead	30,000	Filtration, Reduction, Calcination IV/V	–47	61	NA
11	Ethanolamines 14,000	Triethanolamine Tars	1,120	Centrifugation, Distillation, IV/V	188	77	120

Table IV, continued

Stream[a] No.	Product and Typical Plant Size (metric ton/yr)	Waste Stream Components	Waste Generation[b] (metric ton/yr)	Alternative Treatment Processes	Alternative Treatment Cost[c] ($/metric ton waste)	Chemical Landfill Cost ($/metric ton waste)	Land Incineration Cost ($/metric ton waste)
			Organic Chemicals Industry				
12	Furfural 35,000	Sulfuric Acid Tars Polymers	19,600	Sedimentation, Hydrolysis, Composting III/IV	50	76	32
13	Furfural 35,000	Fines & Particulates from Stripped Hulls	350	Distillation, Hydrolysis, Composting III/IV/V	Combined with Stream No. 12		143
14	Fluorocarbon 80,000	Antimony Pentachloride Carbon Tetrachloride Trichloro-fluoromethane Organics	18	Distillation, Reduction IV or Distillation, IV	5560 470	117	NA
15	Chlorotoluene 15,000	Benzylchloride Benzotrichloride	15	None		156	NA
16	Chlorobenzene 32,000	Polychlorinated Aromatic Resinous Material	1,400	None		70	97
			Pesticides Industry				

17	Atrazine 20,000	Water Sodium Chloride Insoluble Residues Caustic Cyanuric Acid	224,600	Neutralization, Ozonation, Aerated Lagoon, Evaporation II/III/V	4.60	NA	NA
18	Trifluraline 10,000	Spent Carbon Fluoroaromatics Intermediates Solvents	1,150	Grinding, Solvent Extraction, Centrifugation, Distillation, Composting III/IV/V	400	326	123
19	Malathion 14,000	Filter Aid Toluene Insoluble Residues Dimethyl Dithiophosphoric Acid	1,826	Hydrolysis, Steam Stripping, Sedimentation, Composting, Aerated Lagoon III/IV/V	93	326	91
20	Malathion 14,000	Malathion Toluene Impurities Sodium Hydroxide	14,350 (wet) 350 (dry)	Sedimentation, Resin Adsorption, Distillation III/V	–0.38	76	30 (wet) 1240 (dry)
21	Parathion 20,000	Diethylthio- phosphoric Acid	2,300	Sedimentation, Ultrafiltration, Filtration, Composting II/IV/V	73	70	69

[a] See Table III.
[b] The yearly waste generation quantities are for hypothetical, typical plants of the sizes shown. Wastes are nonaqueous unless otherwise indicated.
[c] Includes credit for material recovery where applicable.
[d] NA = not applicable.

Table V. Categorization of Processes

Category No.	Description
I	Process is not applicable in a useful way to wastes of interest to this program
II	Process might work in 5-10 years, but needs research effort first
III	Process appears useful for hazardous wastes, but needs development work
IV	Process is developed but not commonly used for hazardous wastes
V	Process will be common to most industrial waste processors

Inspection of Table VII results in the following observations:

- Alternative treatment costs ranged from \$6.90 to \$3170/metric ton (fourth-quarter 1976 dollars).
- Chemical landfill costs ranged from \$2.60 to \$236/metric ton.
- Chemical landfill costs were less than the alternative treatment costs for all waste streams except one–Stream 16B. However, chemical landfill costs are based on pre-RCRA requirements and can be expected to increase when the RCRA regulations are established.
- Eight of the streams have a water content of 50% or greater.
- Ten of the streams contain heavy metals. This raises questions both on landfill acceptability and ocean incineration. However, ocean incineration may be acceptable if the stream is sufficiently dilute, particularly if the incineration process could prevent exit of the heavy metals with the exhaust gases. In that event, ocean incineration might be preferable to landfill where the heavy metal waste would always be a potential threat to the environment.
- Only five of the streams contain organic material. This means that they are not very combustible and, combined with a high water content, increase the cost of incineration.
- Due to the inorganic nature of the waste stream constituents, much of it will issue as particulates and droplets from the incineration process rather than as a vapor and a portion will remain as residue in the incinerator.

Metal Smelting and Refining

The metal smelting and refining industries include iron and steel, ferroalloys and nonferrous metals–copper, lead, zinc, aluminum, antimony and titanium. The EPA assessment study [15] identified 36 hazardous waste streams; 34 were analyzed for alternative treatments in the EPA alternatives study [16]. The waste streams are defined in Table VIII.

The results of the alternatives study [16] for metal smelting and refining are summarized in Table IX.

Inspection of Table IX results in the following observations:

- Alternative treatment costs ranged from a return of \$18.40 to \$379.30/metric ton due to chemical recovery (fourth-quarter 1976 dollars).
- Chemical landfill costs ranged from \$7.50 to \$98.50/metric ton.
- All but two of the waste streams contain heavy metals (streams 20 and 21). However, these two streams are toxic because they contain fluorine, and one of them (stream 21) also contains cyanide.

Table VI. Hazardous Waste Streams–Inorganic Chemicals Industry

Waste Stream No.	Description
1	Mercury contaminated brine purification muds–mercury cell process, chloralkali production
2	Mercury-rich wastes from treatments and cleanings–mercury cell process, chloralkali production
3	Chlorinated hydrocarbons–diaphragm cell process, chloralkali production
4	Asbestos separator wastes–diaphragm cell process, chloralkali production
5	Lead-containing wastes–diaphragm cell process, chloralkali production
6	Metallic sodium and calcium filter cake–Down's cell process, metallic sodium production
7	Sludges from wastewater treatment–chloride process, titanium dioxide production
8	Sludges from wastewater treatment–chrome color and inorganic pigment production
9	Fluoride-containing gypsum waste–hydrofluoric acid production
10	Arsenic-containing sludges–boric acid production
11	Fluoride-containing wastewater treatment sludge–aluminum fluoride production
12	Antimony waste stockpile–antimony production
13	Fluoride wastes–sodium silicofluoride production
14	Chromate contaminated wastes–chromate production
15	Nickel wastes from wastewater treatment–nickel sulfate production
16A	Fluoride-bearing sludge from phosphate rock calcining kiln and electric furnace–furnace process, phosphorus production
16B	Phossy water from phosphorus condenser–furnace process, phosphorus production
17	Arsenic and phosphorus wastes–phosphorus pentasulfide production
18	Arsenic trichloride waste–phosphorus trichloride production

- Except for those waste streams which are dry, all remaining waste streams contain a water content of at least 60%. This means 10 waste streams with no water and 24 waste streams ranging from 60-90% water.
- Because of the nature of the wastes–heavy metals and high water content–land incineration was not considered a viable option for the industry by the alternatives study [16] and was not further evaluated in that study.
- The wastes are essentially incombustible which means that the gaseous effluent would contain a portion of the wastes in the form of particulates and droplets and that a portion would remain as residue in the incinerator.

Table VII. Cost Comparison of (Basis–Fourth

Stream[a] No.	Product and Typical Plant Size (metric ton/yr)	Waste Stream Components	Waste Generation[b] (metric ton/yr)	Alternative Treatment Processes
1	Chloralkali Mercury Cell Plant 91,250	Mercury, Mercury Chloride, Mercury Sulfide, Calcium Carbonate, Magnesium Hydroxide, Sodium Chloride	6753 77% water	Neutralization, Dissolution, Filtration, Sedimentation, Precipitation, IV/V
2	Chloralkali Mercury Cell Plant 91,250	Mercury Sulfide, Graphite	365 50% water	
3	Chloralkali Diaphragm Cell 164,250	Chlorinated Hydrocarbons	73 0% water	
4	Chloralkali Diaphragm Cell 164,250	Asbestos, Carbon, Rubble	799 30% water	Drying, Calcining, IV/V
5	Chloralkali Diaphragm Cell 164,250	Lead, Lead Carbonate, Asbestos	1511 97% water	Flocculation, Filtration, Calcination/ Smelting, IV/V
6	Sodium 51,100	Sodium Sludge, Calcium Sludge, Lime, Sodium Oxide	810 0% water	Electrolysis, Grinding, Drying, III/V
7	Titanium Dioxide 36,500	Chromium Hydroxide, Ore Residue, Coke, Silica, Misc. Hydroxides	1,022,000 95% water	Flocculation, Sedimentation, Filtration, Calcination, IV/V

Alternatives–Inorganic Chemicals Industry
Quarter 1976)

Alternative Treatment Cost[c] ($/metric ton waste)	Chemical Landfill Cost ($/metric ton waste)	Remarks
89	34	Streams 1 and 2 are combined before waste treatment. A second alternative treatment process is available at a cost of $34/metric ton waste but has not operated satisfactorily and is, therefore, a more questionable alternative.
	63	No alternative treatment since incineration is the preferred disposal method.
252 } 144 91 }	63	Smelter residue that is formed during processing of waste stream no. 5 is incorporated into waste stream no. 4. The stream 4 and 5 alternative treatment disposal costs are shown combined for comparison purposes with a combined chemical landfill disposal cost.
3170	100	Alternative treatment cost is for the waste stream as generated which means on a dry basis instead of the usual wet basis since the stream contains no water. Landfill is not considered an acceptable disposal method since the waste is explosive in contact with water. The $100/ metric ton is for waste enclosed in drums and sent to chemical landfill.
6.9	2.6	The $6.9/metric ton is based on the waste stream as it comes from the manufacturing process. The flocculation, sedimentation, and filtration are part of the dewatering process.

Table VII,

Stream[a] No.	Product and Typical Plant Size (metric ton/yr)	Waste Stream Components	Waste Generation[b] (metric ton/yr)	Alternative Treatment Processes
7 (cont'd)	Titanium Dioxide 36,500	Chromium Hydroxide Ore Residue Coke Silica Misc. Hydroxides	204,400 75% water	Calcination IV
8	Chrome Pigments 8,395	Chromium Hydroxide Lead Chromate Lead Hydroxide Zinc Oxide Iron Cyanide Iron Oxide Calcium Sulfate Iron Hydroxide	1095 23% water	Calcination IV
9	Hydrofluoric Acid 23,360	Calcium Fluoride Calcium Sulfate	109,500 20% water	Evaporation Asphalting IV
10	Boric Acid 40,150	Arsenic-Containing Waste Filter Aid	73-183 25% water	
11	Aluminum Fluoride 52,925	Calcium Fluoride Hydrated Lime Calcium Sulfate	12,045 21% water	Evaporation Asphalting IV
12	Antimony			
13	Sodium Silicofluoride 16,425	Calcium Fluoride Silica Salt Hydrated Lime	8,395 70% water	Evaporation Asphalting IV

continued

Alternative Treatment Cost[c] ($/metric ton waste)	Chemical Landfill Cost ($/metric ton waste)	Remarks
31.2	11.5	The $31.2/metric ton is based on the waste stream after it is dewatered from 95% to 75% water content. Dewatering cost (from 95% to 75%) is $0.66/metric ton of the 95% stream. The remaining $31.2/ metric ton cost appears higher since it is based on the smaller quantity of 75% water content waste stream.
342	107	
77	15	The alternative process may not be environmentally acceptable since it does not change the toxic calcium fluoride component. Instead, it incorporates it in asphalt and produces a larger quantity of waste for land disposal (140,264 metric ton/yr) vs the original 109,500 metric ton/yr
		No alternative waste disposal process was developed. Current practice is to drum the waste and dispose of it in a landfill.
74	23	Same as for waste stream No. 9.
		This waste stream is no longer being generated.
75	35	Same as for waste stream No. 9 except that a smaller quantity of waste results than was originally produced by the manufacturing process.

Table VII,

Stream[a] No.	Product and Typical Plant Size (metric ton/yr)	Waste Stream Components	Waste Generation[b] (metric ton/yr)	Alternative Treatment Processes
14	Chromates 66,430	Chromium Hydroxide Chromate Ore Residue including oxides and carbonates of iron, silicon and magnesium plus calcium salts and aluminum hydroxide	73,000 25% water	Calcination IV
15	Nickel Sulfate 3,285	Nickel Hydroxide Filter Aid Organic Material	365 50% water	High Gradient Magnetic Separation (HGMS) II
16A	Phosphorus 49,640	Calcium Fluoride Lime Silicon Dioxide Calcium Phosphate Calcium Sulfate	30,770 40% water	Evaporation Asphalting IV
16B	Phosphorus 49,640	Phosphorus Suspended Solids	6,680 90% water	Flocculation Heating Distillation IV/V
17	Phosphorus Pentasulfide 55,000	Arsenic Sulfide Phosphorus Phosphorus Sulfide Glassy Phosphate Iron Sulfide	119 0% water	
18	Phosphorus Trichloride 58,000	Arsenic Glassy Phosphate Iron Chloride	60 0% water	Distillation V

[a]See Table VI.
[b]Waste quantities and costs are for the streams as generated, including water content.
[c]Includes credit for material recovery where applicable.

continued

Alternative Treatment Cost[c] ($/metric ton waste)	Chemical Landfill Cost ($/metric ton waste)	Remarks
54	14	
212	118	
91		The stream 16A and 16B alternative treatment disposal costs are shown combined for comparison with a combined chemical landfill disposal cost.
77 (combined)	55	
12	178	Phosphorus recovery reduces the alternative treatment cost from $60 to 12/metric ton
	236	No alternative waste disposal process was developed.
1740	389	

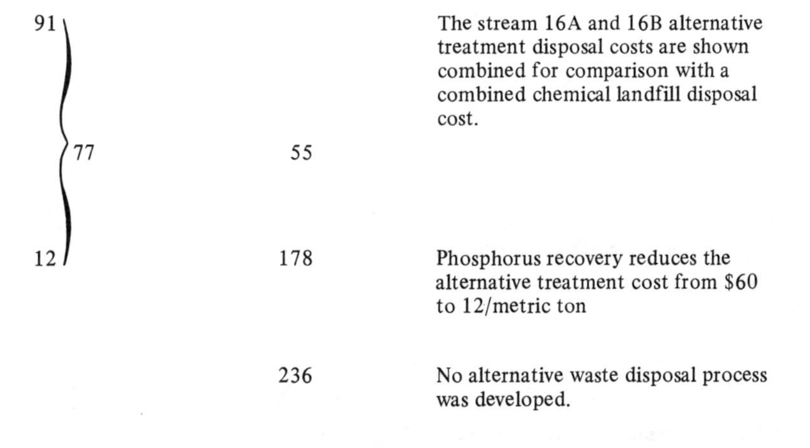

Table VIII. Hazardous Waste Streams–Metals Smelting and Refining Industries

Waste Stream No.	Description
	Ferrous Metal Smelting and Refining
Iron and Steel Coke Production	
1	Ammonia Still Lime Sludge
2	Decanter Tank Tar from Coke Production
Iron and Steel Production	
3	Basic Oxygen Furnace–Wet Emission Control Unit Sludge
4	Open Hearth Furnace–Emission Control Dust
5	Electric Furnace–Wet Emission Control Sludge
6	Rolling Mill Sludge
7A	Cold Rolling Mill–Acid Rinsewater Neutralization Sludge (Sulfuric Acid Rinse)
7B	Cold Rolling Mill–Acid Rinsewater Neutralization Sludge (Hydrochloric Acid)
8A	Cold Rolling Mill–Waste Pickle Liquor (Sulfuric Acid)
8B	Cold Rolling Mill–Waste Pickle Liquor (Hydrochloric Acid)
9A	Galvanizing Mill–Acid Rinsewater Neutralization Sludge (Sulfuric Acid)
9B	Galvanizing Mill–Acid Rinsewater Neutralization Sludge (Hydrochloric Acid)
10	Waste stream number 10 was omitted from this study because it is normally recycled
Ferroalloys	
11	Ferrosilicon Manufacture–Miscellaneous Dust
12A	Ferrochrome Manufacture–Slag
12B	Ferrochrome Manufacture–Dust
12C	Ferrochrome Manufacture–Sludge
13	Silicomanganese Manufacture–Slag and Scrubber Sludge
14	Ferromanganese Manufacture–Slag and Sludge
	Primary Nonferrous Smelting and Refining
15	Copper Smelting–Acid Plant Blowdown Sludge
16	Electrolytic Copper Refining–Mixed Sludge
17	Lead Smelting–Sludge
18	Electrolytic Zinc Manufacture–Sludge

Table VIII, continued

Waste Stream No.	Description
	Primary Nonferrous Smelting and Refining
19A	Pyrometallurgical Zinc Manufacture–Sludges–Primary Gas Cleaning and Acid Plant Blowdown
19B	Pyrometallurgical Zinc Manufacture–Sludges–Retort Gas Scrubber Feed
Aluminum Manufacture	
20	Scrubber Sludges
21	Spent Potliners and Skimmings
22	Shot Blast and Cast Houst Dusts
23	Pyrometallurgical Antimony Manufacture–Blast Furnace Slag
24	Electrolytic Antimony Manufacture–Spent Anolyte Sludge
25	Titanium Manufactur –Chlorinator Condenser Sludge
26	Waste stream number 26, concerning smelter slag from primary tin manufacture, was deleted from this study because of insufficient information
	Secondary Nonferrous Refining
27	Copper Refining–Blast Furnace Slag
28	Lead Refining–SO_2 Scrubwater Sludge
Aluminum Refining	
29	Scrubber Sludge
30	High Salt Slag

Petroleum Refining

The waste streams from the petroleum refining industry were determined under a previous EPA industry assessment study [17] and are summarized in Table X. These streams were then examined in a follow-up EPA study [18] for alternative waste disposal methods which also further characterized certain of the waste stream components.

The waste streams are detailed in Tables XI and XII. The quantities in Table XI are based on a typical refinery with a capacity of 200,000 bbl/day. The solids in Table XI consist primarily of corrosion products, rust and sand with a very small proportion of hazardous components. These hazardous components are shown in Table XII and are expressed in ppm concentration

Table IX. Cost Comparison of Alternatives–Metals Smelting and Refining Industries (Basis–Fourth Quarter 1976)

Stream[a] No.	Product and Typical Plant Size (metric ton/yr)	Waste Stream Components		Waste Generation[b] (metric ton/yr)	Alternative Treatment Processes	Alternative Treatment Cost[c] ($/metric ton waste)	Chemical Landfill Cost ($/metric ton waste)
1	Steel 2,500,000	Chromium Copper Nickel Cyanide Oil & Grease	Zinc Manganese Lead Phenol	2,300 70% water	Disposal V	78.9	78.9
2	Steel 2,500,000	Chromium Copper Nickel Cyanide Oil & Grease	Zinc Manganese Lead Phenol	27,600 70-85% water	Disposal V	64.6	64.6
3	Steel 2,000,000	Iron Manganese Cyanide Copper	Zinc Lead Chromium Nickel	86,500 60% water	Reduction Roasting V		38.8
4	Steel 500,000	Iron Manganese Chromium Nickel	Zinc Lead Copper	6,900 0% water	Reduction Roasting V	7.4	16.7
5	Steel 500,000	Iron Manganese Chromium Nickel	Zinc Lead Copper Fluorine	10,900 60% water	Reduction Roasting V		42.1

6	Steel 1,800,000	Iron Iron Oxide Chromium Copper Manganese	Nickel Lead Zinc Oil & Grease	7,800 60% water	Sintering V	1.5	54.3
7A	Steel 700,000	Calcium Manganese Lead Iron Iron Oxides Phenol	Chromium Nickel Copper Zinc Oil & Grease Cyanide	400 70% water	Dissolution V	6.9	73.9
7B	Steel 700,000			300 90% water	Dissolution V	6.8	98.5
8A	Steel 700,000	Iron Sulfate Sulfuric Acid Chromium Copper	Manganese Nickel Lead Zinc Oil & Grease	78,700 80% water	Precipitation III	43.3	70.4
8B	Steel 700,000	Iron Chloride Hydrochloric Acid Chromium Copper	Manganese Nickel Lead Zinc Oil & Grease	37,500 80% water	Volatilization Reduction Roasting IV	24.8	70.3
9A	Galvanized Steel 125,000	Iron Iron Oxides Calcium Chromium	Nickel Copper Lead Oil & Grease	4,500 70% water	Dissolution V	1.0	57.9
9B	Galvanized Steel 125,000			1,000 70% water	Dissolution V	3.7	88.7

Table IX, continued

Stream[a] No.	Product and Typical Plant Size (metric ton/yr)	Waste Stream Components		Waste Generation[b] (metric ton/yr)	Alternative Treatment Processes	Alternative Treatment Cost[c] ($/metric ton waste)	Chemical Landfill Cost ($/metric ton waste)
10	This waste stream was omitted from this study since it is normally recycled.						
11	Ferrosilicon 40,000	Silica Iron Oxide Ferrosilicon Lime Chromium	Copper Zinc Manganese Nickel Lead Cobalt	13,500 0% water	Disposal	15.9	10.0
12A	Ferrochrome 35,000	Chromium Chromic Oxide Silica Alumina Magnesium Oxide Lime	Iron Oxide Carbon Copper Lead Zinc Manganese	61,300 0% water	Precipitation V	2.9	8.8
12B	Ferrochrome 35,000	Same as 12A but different proportions and amounts.		5,300 0% water	Precipitation V	18.8	9.9
12C	Ferrochrome 35,000	Same as 12A but different proportions and amounts.		13,200 60% water	Precipitation V	13.9	64.6
13	Silicomanganese 40,000	Silica Alumina Lime Magnesium Oxide Zinc	Manganese Oxide Manganese Chromium Lead Copper	24,000 0% water	Reduction Roasting IV	18.8	8.8

14	Ferromanganese 30,000	Zinc Lead Manganese Lime	Copper Chromium Nickel	22,200 60% water	Reduction Roasting IV	18.8	55.5
15	Copper 100,000	Arsenic Cadmium Silicon Copper Mercury Manganese	Nickel Lead Antimony Selenium Zinc	700 60% water	Precipitation V	379.3	84.4
16	Copper 160,000	Copper Lead Cadmium Chromium Mercury	Manganese Nickel Antimony Selenium Zinc	1,100 60% water	Precipitation V	360.4	62.7
17	Lead 110,000	Lead Zinc Cadmium Copper Mercury	Arsenic Chromium Manganese Nickel Antimony Selenium	21,600 70% water	Sintering V	–1.0[d]	64.6
18	Zinc 100,000	Zinc Cadmium Copper Mercury	Chromium Manganese Lead Selenium Lime	8,700 70% water	Precipitation V	–3.5	62.4

Table IX, continued

Stream[a] No.	Product and Typical Plant Size (metric ton/yr)	Waste Stream Components		Waste Generation[b] (metric ton/yr)	Alternative Treatment Processes	Alternative Treatment Cost[c] ($/metric ton waste)	Chemical Landfill Cost ($/metric ton waste)
19A	Zinc 107,000	Zinc Cadmium Copper	Lead Mercury Lime	43,000 70% water	Sintering V	−15.4[d]	59.7
19B	Zinc 107,000	Cadmium Chromium Copper Lead	Selenium Zinc Mercury	2,200 70% water	Centrifuge V	15.3	53.5
20	Aluminum 153,000	Fluorine Aluminum	Carbon Sodium Lime	59,500 70% water	Precipitation Evaporation Dewatering Drying Disposal V (streams 20 and 21)	−18.4 (streams 20 and 21)	52.2
21	Aluminum 153,000	Fluorine Aluminum	Carbon Sodium Cyanide	9,000 0% water			9.0
22	Aluminum 153,000	Iron Carbon Fluorine Copper Chromium	Aluminum Lead Zinc Manganese Nickel	1,100 0% water	Precipitation V	75.3	25.5

23	Antimony 2,700	Silicon Dioxide Iron Oxide Alumina Antimony Oxide	Lead Copper Zinc Arsenic Cadmium	7,700 0% water	Precipitation V	18.4	10.9
24	Antimony 900	Antimony Arsenic Lead Copper	Zinc Nickel Cadmium Chromium	600 70% water	Disposal V	55.1	67.9
25	Titanium 7,600	Vanadium Chromium Titanium	Zinc Chlorine Carbon	6,300 60% water	Centrifuge Dewatering Recycling III	–14.9	67.3
26	This waste stream, smelter slag from primary tin manufacture, was omitted from the study because of insufficient information.						
27	Copper 10,000	Zinc Cadmium Chromium Copper	Manganese Lead Antimony Tin	3,500 0% water	Precipitation V	37.9	13.6
28	Lead 10,000	Cadmium Chromium Copper Manganese Calcium Sulfate	Nickel Lead Antimony Zinc Calcium Sulfite	1,500 70% water	Precipitation V	25.6	75.0

Table IX, continued

Stream[a] No.	Product and Typical Plant Size (metric ton/yr)	Waste Stream Components		Waste Generation[b] (metric ton/yr)	Alternative Treatment Processes	Alternative Treatment Cost[c] ($/metric ton waste)	Chemical Landfill Cost ($/metric ton waste)
29	Secondary Aluminum 20,000	Fluoride Chloride Sodium Chromium	Copper Lead Zinc	5,000 70% water	Centrifuge Dewatering V	16.6	43.5
30	Secondary Aluminum 10,000	Aluminum Alumina Fluxing Salts Chromium	Copper Manganese Nickel Lead Zinc	14,000 0% water	Crushing Screening Dewatering Drying Dissolution Evaporation IV	47.9	7.5

[a]See Table VIII.
[b]The waste quantities and costs are on the wastes as generated, including water, where that is part of the waste stream.
[c]Includes credit for material recovery where applicable.
[d]A minus sign indicates a cost credit.

Table X. Hazardous Waste Streams–Petroleum Refining Industry [18]

Stream No.	Stream Designation
1	Slop Oil Emulsion Solids
2	Storm Water Silt
3	Exchanger Bundle Cleaning Sludge
4	API Separator Sludge
5	Nonleaded Tank Bottoms
6	Crude Tank Bottoms
7	Leaded Tank Bottoms
8	Dissolved Air Flotation Float
9	Kerosene Filter Clays
10	Lube Oil Filter Clays
11	Neutralized HF Alkylation Sludge
12	Spent Lime from Feedwater Treatment
13	Waste Bio Sludge
14	Once-Through Cooling Water Sludge
15	Cooling Tower Sludge
16	FCC Catalyst Fines
17	Coke Fines

based on the quantity of solids. Alternative waste disposal methods and preliminary costs were developed for 15 of the 17 streams and the results are summarized in Table XIII. Note that streams 1-10 were combined for treatment instead of being treated separately; likewise, streams 12-15. Also, only several alternative treatment processes were employed.

Waste Quantities

Establishment of toxic waste quantities for the U.S. was based on data for chemical and waste production in the previously referenced EPA studies for the four industries of interest (organic/pesticide, inorganic, petroleum refining, metal smelting/refining). The quantities were then extrapolated to 1977, 1983 and 1989 based on chemical growth rate and the impact of environmental laws. Foreign toxic waste quantities were established by approximating chemical production and then applying waste factors that were specific for each country involved.

Chemical Growth Rate

Extrapolations for future chemical production were based on the historical relationship between chemical production and gross national product (GNP).

The average annual growth rate, in five-year groupings, for the last 25 years for both the chemical industry and the GNP is shown in Table XIV, including the ratio between the two. The basic data for the chemical growth index and

Table XI. Composition of Waste Streams from a Typical Refinery [18] (32,000 m^3/day)

Stream No.	Stream Designation	Total Quantity (metric ton/yr)	Breakdown (metric ton/yr)			Oil in Waste Stream (%)
			Oil	Water	Solids	
1	Slop Oil Emulsion Solids	2,300	1,104	920	276	48.0
2	Silt from Storm Water Runoff	562	22	140	400	3.9
3	Exchanger Bundle	182	20	96	66	10.7
4	API Separator Sludge	1,803	407	956	440	22.6
5	Nonleaded Tank Bottoms	5,083	3,263	900	920	64.2
6	Crude Tank Bottoms	29	14	4	11	47.4
7	Leaded Tank Bottoms	29	6	0	23	20.0
8	Dissolved Air Flotation Skimings	13,230	1,650	10,850	730	12.5
9	Kerosene Filter Clays	203	7	10	186	3.4
10	Lube Oil Filter Clays	2,636	577	132	1,927	21.9
11	Neutralized HF Alkylation Sludge	823	57	446	320	6.9
12	Spent Lime from Boiler Feedwater	14,200	45	8,378	5,777	0.32
13	Waste Bio Sludge	1,258	4	1,094	160	0.28
14	Once-Through Cool-ling Water Sludge	3,380	15	845	2,520	0.43
15	Cooling Tower Sludge	24	0	18	6	0.44
16	FCC Catalyst Fines	371	1	0	370	0.21
17	Coke Fines	2	0	0	1.6	0

the GNP were obtained from "Business Statistics 1975" [19], as updated by the monthly publication, "Survey of Current Business" [20].

Although the average value of the ratio over the past 25 years is approximately 2.4 from the above table, inspection of the ratio shows that it is in a declining trend, i.e., the rate of increase in chemical growth is decreasing relative to GNP. Therefore, instead of projecting the 2.4 ratio into the future, the individual ratios were plotted in Figure 1, the least-squares regression curve was calculated and the decreasing ratios were used for establishing the estimated chemical production quantities for 1977, 1983 and 1989. The decreasing ratios and chemical growth rates are shown in Table XV. The projected real GNP growth rates (based on constant dollars) were based on an article prepared by John Kendrick, Chief Economist for the Department of Commerce, for the publication "U.S. Industrial Outlook" [21].

Waste Generation Factors

There are a number of factors which will influence potential ocean waste incineration quantities. These are:

1. Chemical growth rate, which has been projected as 8.0% for 1977-81, 6.5% for 1982-86, and 5.2% for 1987-91 (see Table XV).

2. Decreased percentage of waste discharged to the nation's waterways or to air in response to the Clean Water [22] and Clean Air [23] Acts. These decreased wastes will essentially be diverted to land disposal and therefore do not represent a decrease in total waste but rather a corresponding increase in land waste. Based on implementation requirements and the extent of industry compliance to date, we have estimated a percentage decrease in discharge to water of 10% by 1977, 50% by 1983 and 80% by 1989.

Data from the EPA's National Air Bank in Research Triangle Park, NC, give an indication of the impact of the Clean Air Act on pollution quantities

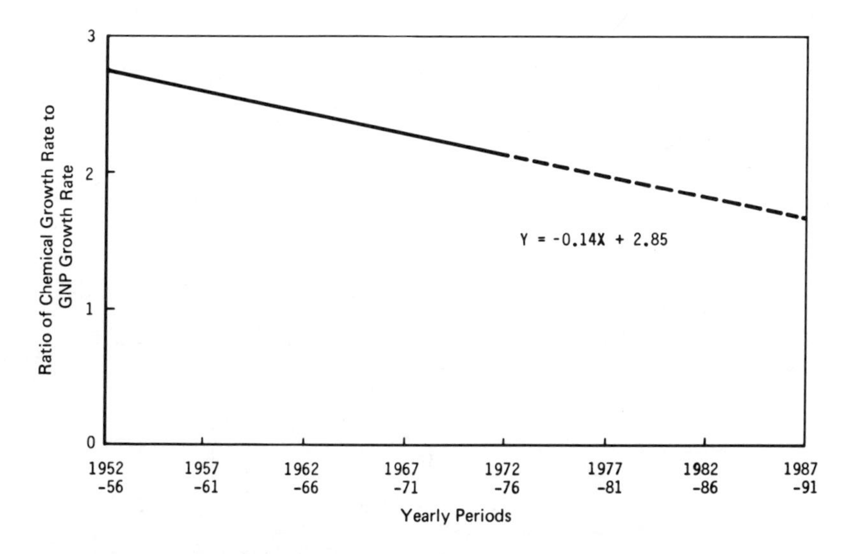

Figure 1. Relationship of chemical growth to GNP.

Table XII. Waste Stream Hazardous Component Concentration [18] (ppm by weight)

Component	Stream #1	Stream #2	Stream #3	Stream #4	Stream #5	Stream #6	Stream #7	Stream #8
Phenols	15.0	7.5	13.3	13.6	1.8	15.8	126	6.5
Cyanide	0.001	1.72	1.7	0.001	7.4	0.0012	0.0009	0.28
Selenium	1.0	1.7	27.2	0.001	12.0	0.03	6.95	2.0
Arsenic	7.4	5.5	10.6	6.2	0.007	21.1	294.0	2.0
Mercury	0.59	0.3	1.9	0.4	0.43	0.48	0.57	0.27
Beryllium	0.0025	0.0019	0.2	0.0025	0.26	0.0026	0.0025	0.0025
Vanadium	25.0	69.0	25.0	9.8	21.9	17.4	5.4	0.05
Chromium	525.0	354.0	311.0	253.0	12.9	19.4	11.4	140.0
Cobalt	10.0	11.2	1.6	5.7	7.1	14.8	49.0	2.0
Nickel	50.0	85.0	116.0	0.913	26.7	16.2	314.0	0.025
Copper	48.0	28.3	71.0	18.6	85.1	65.4	141.0	7.0
Zinc	250.0	230.0	194.0	298.0	285.0	145.0	9,090.0	85.0
Silver	0.4	0.5	0.005	0.45	0.6	0.19	0.88	0.25
Cadmium	0.19	0.22	1.3	0.42	0.33	0.31	6.3	0.005
Lead	28.1	53.3	78.0	26.0	24.7	18.9	790.0	7.5
Molybdenum	5.0	6.9	6.5	5.0	9.2	6.3	59.0	0.05
Ammonia Salts	8.0	1.0	8.0	6.5	0.2	2.0	—	9.0
Benz[a]Pyrene	0.003	1.27	2.2	0.004	0.6	0.11	0.21	0.002
TOTAL	973.7	857.4	869.5	644.6	496.2	343.4	10,844.7	261.9

Component	Stream #9	Stream #10	Stream #11	Stream #12	Stream #13	Stream #14	Stream #15	Stream #16	Stream #17
Phenols	2.6	3.2	8.9	2.1	4.5	2.0	3.5	2.1	2.0
Cyanide	0.0005	0.12	23.1	0.001	0.001	0.11	0.1	0.12	0.001
Selenium	9.6	1.1	7.1	0.001	0.01	1.6	0.015	0.01	0.01
Arsenic	2.2	0.7	2.3	0.1	3.8	5.4	8.2	1.0	2.0
Mercury	0.0004	0.19	0.07	0.04	0.18	0.51	0.09	0.0004	0.04
Beryllium	0.02	0.26	0.07	0.0012	0.0013	0.24	0.0013	0.5	0.005
Vanadium	35.0	0.33	2.6	0.05	0.05	39.0	7.8	240.0	455.0
Chromium	2.88	23.2	2.8	2.2	300.0	6.30	554.0	71.0	0.02
Cobalt	0.62	3.2	0.36	0.005	0.02	6.3	1.3	6.6	4.0
Nickel	0.04	11.1	55.2	2.5	0.025	35.0	6.8	241.0	580.0
Copper	11.4	4.3	14.3	3.8	9.5	153.0	50.0	17.5	4.0
Zinc	25.0	57.8	9.1	15.0	122.0	125.0	675.0	39.0	14.0
Silver	0.5	0.5	0.19	0.5	0.3	0.5	0.28	1.8	0.01
Cadmium	0.25	0.76	0.07	0.003	0.3	0.003	0.3	0.003	1.0
Lead	8.3	1.28	7.1	3.8	5.0	57.0	38.0	50.0	13.0
Molybdenum	0.025	0.38	0.0025	0.0025	2.5	2.5	1.1	6.3	0.1
Ammonia Salts	0.01	2.0	0.002	0.015	21.0	6.0	6.3	0.01	0.7
Benz[a]Pyrene	1.8	0.1	0.002	0.002	0.003	0.004	0.004	0.005	0.002
TOTAL	100.2	110.5	133.3	30.2	469.2	497.2	1352.8	679.9	1075.9

Table XIII. Alternative Waste Disposal Methods–Petroleum Refining Industry [18]

Stream No.	Alternative Treatment Process	Alternative Treatment Cost[a] ($/metric ton dry waste[b])	Chemical Landfill Cost ($/metric ton dry waste)
1-10	Phase Separation	36.13	86.61
11	Phase Separation, Metal Precipitation	123.98	41.94
12-15	Calcination	26.59	34.38
16-17	None–Dispose to Landfill		

[a]Includes credit for material recovery where applicable.
[b]Dry waste means cost expressed in $/metric ton of nonaqueous waste.

Table XIV. Average Annual Growth Rates (%)

Period No.	Period	Avg. Annual Chem. Growth	Avg. Annual Real[a] GNP Growth	Ratio of Chem. Growth Rate to GNP Growth Rate
1	1952-56	7.66	3.1	2.47
2	1957-61	6.88	2.2	3.13
3	1962-66	10.84	5.8	1.87
4	1967-71	6.5	2.38	2.73
5	1972-76	6.12	3.1	1.97

[a]In constant dollars.

Table XV. Projected Ratios of Chemical Growth Rate to GNP Growth Rate

Period	Ratio	GNP Growth Rate	Chemical Growth Rate[a]
1977-81	2.01	1976-80 = 4.0%	8.0%
1982-86	1.87	1981-85 = 3.5%	6.5%
1987-91	1.73	1986-90 = 3.0%	5.2%

[a]Chemical growth rate = growth rate × ratio.

[24]. Total emitted airborne waste for particulates, SO_X, NO_X and hydrocarbons has remained approximately constant from the chemical and petroleum refinery industries since 1970–the year the law was enacted. This means that the increasing quantities that would have been airborne each year without the law have essentially been diverted to land disposal. These quantities can be approximated for 1977 by assuming that the airborne waste quantities would have increased at the same rate as chemical growth. For 1983 and 1989 we have estimated that the law would become increasingly effective and, to be consistent with waterborne waste, have applied a 50% decrease for 1983 and 80% for 1989.

3. The impact of RCRA. Although the regulations which implement RCRA have not yet been promulgated (estimated effective data is July 1980 [25]) the emphasis of the law and the draft regulations is such that stricter land disposal requirements can be anticipated. This means greater expense for land disposal and a corresponding increase in the quantity of waste for which ocean incineration becomes economically competitive. Although no firm quantitative values can be calculated at this time, an increase in ocean waste quantities has been estimated as 10% of the waste that would have remained for land disposal without RCRA.

4. Technological advancement. The continuing development of new and improved chemical and manufacturing processes is beginning to include processes which also require less energy and produce less waste. This trend should continue at an accelerating rate, although, again, no firm, quantitative values can be calculated. To recognize this factor we have assumed no impact in 1977, a 2% decrease in waste for 1983, and a 5% decrease for 1989. Examples of such new technology are Union Carbide's vastly simplified low-density polyethylene process [26,27], the new Low Pressure Oxo process for converting propylene to butyraldehyde (Union Carbide-Davy Powergas-Johnson Matthey) [28], and Exxon's Flexicoking process for producing lighter hydrocarbon liquids and coke gas from heavy oils [28].

5. Decreased industrial expansion. Strict environmental requirements are apparently causing delays in and even cancellation of some new industrial plants or planned expansions. It is difficult to establish a quantitative value for this factor, but we have estimated a 1% decrease in waste to recognize this influence. Examples are cancellation of Dow's planned petrochemical complex near San Francisco early in 1977; Phillips Petroleum Company's delay in expanding its Sweeng, TX, refinery [29]; and cancellation of some oil industry expansion plans [30].

These factors are summarized in Table XVI.

Organic Chemicals and Pesticides–United States

Organic chemical and pesticide protection, toxic waste quantities and waste factors were established for 1973, 1977, 1983 and 1989 based on the prior EPA study [11], U.S. International Trade Commission data [31,32], and Table XV extrapolation factors. The total waste quantities from the

Table XVI. Factors Influencing Waste Quantities

No.	Item	1977	1977-81	1983	1982-86	1989	1987-91	Remarks
1	Production Growth Rate, Yearly		8%		6.5%		5.2%	From Table XV
2	Clean Water Act	10%		50%		80%		% of water waste diverted to land disposal
3	Clean Air Act	Same as prod. growth rate		50%		80%		% of air waste diverted to land disposal
4	RCRA	0		10%		10%		Additional % of land waste available for ocean incineration
5	Technology	0		–2%		–5%		Decreased % of land waste
6	Industrial Expansion		–1%		–1%		–1%	Decreased % of production growth
7	U.S. Ocean Incin. Waste Quantities (Economic Basis)	346,000 metric tons		1,081,000 metric tons		1,524,000 metric tons		U.S. Gulf and East Coasts only. 1977 is Gulf Coast only.

study [11] were converted to toxic waste quantities by evaluating the characteristics of the waste stream constituents.

Existing chemical production and waste quantities as of 1973 were 188,552 million pounds of product and 4704 million pounds of toxic land waste for those two industries. The overall waste factor was 0.025 lb toxic waste/lb product. Toxic waste factors from 1973 to 1989 are shown in Table XVII, projected chemical production is shown in Table XVIII, and the corresponding projected waste stream quantities are shown in Table XIX.

Table XVII. Toxic Waste Factors–Organic Chemical and Pesticide Industries, 1973-1989

	Toxic Waste Factors (lb/lb)			
Category	**1973**	**1977**	**1983**	**1989**
Organic Chemicals	0.024	0.025	0.029	0.032
Pesticides	0.088	0.090	0.097	0.100
TOTAL	0.025	0.026	0.031	0.033

Table XVIII. Projected Organic Chemical and Pesticide Production

Year	Total Production (million lb)	Organic Chemicals (million lb)	Pesticides (million lb)
1977	210,232	205,838	4,394
1983	323,055	316,303	6,752
1989	452,826	443,362	9,464

Table XIX. Projected Toxic Waste Stream Quantities–Organic Chemical and Pesticide Industries, 1977-1989

Year	Total Toxic Waste Stream (million lb)	Organic Chemicals (million lb)	Pesticides (million lb)
1977	5,466	5,071	395
1983	10,015	9,360	655
1989	14,943	13,997	946

As can be seen, toxic waste stream quantities for the organic chemical and pesticide industries increase from 4704 million lb in 1973 to 14,943 million lb in 1989 for an overall estimated increase of 218%. Pesticides represent 8.0% of the wastes in 1973 and 6.3% in 1989. The portions of these wastes that are potential candidates for ocean incineration are discussed below.

Criteria for Ocean Incineration. Assuming environmental acceptability, the choice among alternative waste disposal methods will almost always be based on cost considerations. Ocean incineration will be a viable alternative only if it is economically competitive. Therefore, all hazardous waste streams with land disposal costs estimated at less than $30/metric ton in the EPA alternatives studies were eliminated from consideration.

Since ocean incineration costs increase as wastewater content increases, all wastes with a water content exceeding 50% were also eliminated.

Realistically, all of the waste which could be incinerated at sea will not be so incinerated for various reasons, such as:

- too far inland so not economically transportable to the seacoast;
- located near a seacoast which has insufficient quantities available to support an incineration ship;
- the waste can be combined with other wastes for more economical land-based disposal; or
- the waste generating facility has already made other arrangements for waste disposal and does not wish to change.

For the above reasons and to be conservative, the net amount of waste available for ocean incineration was based on alternative disposal cost ($30/ton or less), water content (50% or less), geographical distribution of the waste and then only half of that amount.

The geographical distribution of the toxic wastes generated by the organic chemical industry was determined on a percentage basis by state for all states producing 5000 metric tons or more of waste per year, 1977 basis. This represented over 99% of the waste production and consisted of 19 states. The results are shown in Table XX. Texas was responsible for the largest amount of toxic organic chemical waste with 44.96% of the nation's total. This was 3.65 times the quantity generated by the second state, Louisiana, with 12.31% of the nation's total. The third state was Tennessee with 10.34% which means that the top three states generate 2/3 of the nation's total (67.61%).

Table XXI shows a geographical tabulation of the wastes from that portion of the pesticide industry (SIC 28694) which produces the toxic waste streams that were analyzed in the EPA assessment study [11]. It includes each state generating over 1% of those wastes as of 1977, dry basis, and consists of 15 states which are responsible for 95.5% of the wastes.

U.S. Gulf Coast: Location of the states along the Gulf Coast and on the Mississippi and Ohio Rivers was also examined. Assuming that the wastes can be transported down the rivers to the Gulf, this means that approximately 75% (76.1%, to be exact) of the nation's toxic organic wastes and 66.8% of the pesticide wastes, 1977 basis, could be available on the Gulf Coast for

Table XX. Organic Chemical Toxic Wastes, 1977 (≥5000 metric ton/yr)

State	% of Total	Rank
Alabama[a]	0.87	13
California	0.49	16
Delaware	1.12	11
Florida[a]	1.35	9
Georgia[a]	0.44	18
Illinois[a]	0.58	15
Kentucky[a]	4.78	6
Louisiana[a]	12.31	2
Michigan	1.02	12
New Jersey	2.17	7
New York	0.28	19
North Carolina	1.43	8
Ohio[a]	0.45	17
Pennsylvania[b]	5.49	5
Puerto Rico	0.81	14
South Carolina	1.13	10
Tennessee[a]	10.34	3
Texas[a]	44.96	1
West Virginia[b]	9.21	4
	99.23%	

[a]Has access to the Gulf Coast.
[b]Pennsylvania and West Virginia areas west of the Appalachian Mountains can ship wastes more economically to the Gulf Coast via the Ohio and Mississippi Rivers than to the East Coast, but were not included in the Gulf Coast totals, to be conservative.

incineration ship pickup and disposition. Tables XX and XXI indicate the states with such Gulf Coast access. Pennsylvania and West Virginia were not included as states with Gulf access since their long distance from the Gulf would mean a more expensive transportation charge.

Applying the availability criteria to the waste quantities in Table XIX yields the net ocean incineration quantities for the Gulf Coast as shown in Table XXII. Net organic chemical wastes are 245,000 metric tons in 1977 increasing to 851,000 by 1989.

U.S. East Coast: The U.S. East Coast wastes are primarily generated by four states–Delaware, New Jersey, portions of Pennsylvania and portions of West Virginia. "Portions of Pennsylvania and West Virginia" refers to the fact that the areas of those two states east of the Appalachian Mountains will more economically feed to the East Coast (e.g., via rail and highway or down the Delaware River) whereas areas west of the Appalachians such as Pittsburgh will more economically feed to the Gulf Coast (e.g., down the Ohio and Mississippi Rivers). The other Atlantic or near-Atlantic Coast states generate an insignificant amount of waste. As a result, only 9.2% of the potential U.S.

Table XXI. Geographic Location of Toxic Waste, Dry Bases–Pesticides (SIC 28694), 1977

State	% of Total	Rank
Alabama[a]	20.1	1
Arkansas	1.8	13
California	8.8	5
Illinois[a]	2.8	11
Iowa[a]	2.9	10
Louisiana[a]	12.3	2
Michigan	4.1	9
Mississippi[a]	4.9	6
Missouri[a]	4.3	8
New Jersey	11.8	4
Ohio[a]	1.3	15
Tennessee[a]	4.5	7
Texas[a]	11.9	3
Washington	1.3	14
Wisconsin	2.7	12
TOTAL	95.5%	

[a]Has access to the Gulf Coast.

Table XXII. U.S. Net Ocean Incineration Quantities–Organic Chemicals and Pesticides, Gulf Coast (10^3 metric tons)

Category	1977	1983	1989
Organic Chemicals	245.2	569.1	851.0
Pesticides	10.9	26.1	37.7
TOTAL	356.1	595.2	888.7

organic waste and 11.8% of the potential U.S. pesticide waste can be considered for the East Coast. This yields net ocean incineration quantities for the East Coast as shown in Table XXIII.

Inorganic Chemicals–United States

A similar approach with the inorganic chemicals as for the organic chemicals yielded the total generated waste quantities shown in Table XXIV but the net ocean incineration quantities for the Gulf Coast shown in Table XXV.

Table XXIII. U.S. Net Ocean Incineration Quantities–Organic Chemicals and Pesticides, East Coast (10^3 metric tons)

Category	1977	1983	1989
Organic Chemicals	29.6	68.8	102.9
Pesticides	1.9	4.6	6.7
TOTAL	31.5	73.4	109.6

Table XXIV. Land Disposal of Hazardous Wastes–Inorganic Chemicals, 1974

Industry Category		Amount, Dry/Wet Basis (metric ton/yr)	No. & Type of Plants Land-Disposing
SIC 2812 Alkalies & Chlorine			
SIC 28121	Chlorine	57,000	38 Diaphragm Cell 28 Mercury Cell 5 Down's Cell
TOTAL[a]		57,000/110,000	
SIC 2816 Inorganic Pigments			
SIC 28161	Titanium Dioxide Pigment	150,000	8 Chloride Process
SIC 28162	White Pigments	240	1 Antimony Oxide
SIC 28163	Chrome Colors	8,000	8 Chrome Pigments or Chrome Pigments/ Iron Blues Complexes
TOTAL[a]		160,000/350,000	
SIC 2819 Inorganic Chemicals, NEC			
SIC 28194	Inorganic Acids	1,400,000	12 Hydrofluoric Acid
SIC 28196	Aluminum Compounds	32,000	6 Aluminum Fluoride
SIC 28197	K & Na Compounds	4,400	3 Sodium Silicofluoride
SIC 28199	Others, NEC	350,000	3 Chromates 4 Nickel Sulfate 10 Phosphorus 7 Phosphorus Pentasulfide 5 Phosphorus Trichloride
TOTAL[a]		1,800,000/2,900,000	

[a]Rounded off to two significant figures.

Table XXV. Net Ocean Incineration Quantities–Inorganic Chemical Wastes (Wet Basis), Gulf Coast (10^3 metric tons)

Year	Economic Quantity
1977	37.7
1983	252.8
1989	327.1

Petroleum Refining Industry–United States

The net ocean incineration quantities for the petroleum refining industry, Gulf Coast, are shown in Table XXVI.

Metal Smelting and Refining–United States

None of the waste streams from the metal smelting and refining industry met the criteria for ocean incineration, i.e., water content was too high and alternative land disposal costs were too low. Therefore, these wastes were not considered further in the study.

Overview–United States

The geographic distribution of the wastes from the industries considered is shown in Table XXVII.

Based on the geographical distribution of the waste and on the assumption that only half of the potential waste available to the coast will be assigned to ocean incineration, Table XXVIII shows the net ocean incineration quantities for the Gulf and East Coasts for 1977, 1983 and 1989. The totals are shown as economic (including the inorganics) and environmental (excluding the inorganics). The inorganic wastes which meet the criteria of low water content and expensive land disposal costs almost invariably have heavy metal content. This means that EPA will not approve their ocean incineration unless the incineration system prevents exit of the heavy metals except in trace quantities.

Table XXVI. U.S. Net Ocean Incineration Quantities–Petroleum Refinery Wastes, Gulf Coast (10^3 metric tons)

Year	Wet Basis
1977	52.3
1983	123.0
1989	150.6

Table XXVII. Chemical Waste Distribution, 1977

State	Organic Chem.		Pesticides		Inorganic Chem.		Petro. Refining	
	% of Total	Rank	% of Total	Rank	% of Total	Rank	% of Total	Rank
Alabama[a]			20.1	1	1.1	11		
Arkansas[a]			1.8	13				
California			8.8	5			12.8	2
Delaware	1.1	11					1.2	13
Florida[a]	1.4	9			1.1	12		
Idaho					5.2	4		
Illinois[a]			2.8	11	3.2	8	7.1	4
Indiana[a]							3.8	6
Iowa[a]			2.9	10				
Kansas[a]							3.6	9
Kentucky[a]	4.8	6			5.8	3		
Louisiana[a]	12.3	2	12.3	2	24.6	2	12.3	3
Michigan	1.0	12	4.1	9				
Minnesota							1.1	14
Mississippi[a]			4.9	6			1.6	12
Missouri[a]			4.3	8				
New Jersey	2.2	7	11.8	4	2.8	9	3.8	7
North Carolina	1.4	8						
Ohio[a]			1.3	15	4.0	7	3.8	8
Oklahoma[a]							3.6	10
Pennsylvania[b]	5.5	5			1.1	10	5.9	5
South Carolina	1.1	10						
Tennessee[a]	10.3	3	4.5	7	5.0	5		
Texas[a]	45.0	1	11.9	3	39.0	1	26.9	1
Washington			1.3	14			2.3	11
West Virginia[b]	9.2	4			4.6	6		
Wisconsin			2.7	12				
Wyoming							1.1	15
TOTAL	95.3%		95.5%		97.5%		90.9%	
Gulf Coast access	73.8%		66.8%		83.8%		62.7%	

[a]Gulf Coast access.
[b]Gulf Coast access for area west of the Allegheney Mountains.

Foreign Waste Accumulation

Net ocean incineration waste quantities for major world areas were established based on GNP and chemical industry growth rate similar to the method used for the U.S., including use of Table XV-type factors specific to the countries involved. The initial total waste quantities were based on the production ratio of key chemicals in the foreign country to that in the U.S. For

Table XXVIII. U.S. Net Ocean Incineration Waste (10^3 metric tons)

	1977	1983	1989
Gulf Coast			
Organic Chem.	245	569	851
Pesticides	11	26	38
Inorganic Chem.	38	253	327
Petroleum Refining	52	123	151
Eco. Total[a]	346	971	1367
Env. Total	308	718	1040
East Coast			
Organic Chem.	30	69	103
Pesticides	2	5	7
Inorganic Chem.	2	15	19
Petroleum	9	21	28
Eco. Total	43	110	157
Env. Total	41	95	138
U.S. Grand Totals			
Economically Avail.	389	1081	1524
Environmentally Avail.	349	813	1178

[a]Economic totals are the complete totals, environmental totals are without the inorganic chemicals.

example, Canadian production of benzene plus ethylene in 1977 [33] was 5.9% of U.S. production. Therefore, waste produced by the Canadian organic chemical industry in 1977 would be approximately 5.9% of the U.S. amount.

The foreign countries for which waste quantities were estimated and projected consisted of the following:

	Western Europe	
Brazil	Austria	Italy
Canada	Belgium	Luxembourg
Japan	Denmark	The Netherlands
Mexico	England	Portugal
	Finland	Spain
	France	West Germany
	Ireland	

Summarized waste quantities for these countries plus the United States are shown in Table XXIX.

Table XXIX. Major World Net Ocean Incineration Quantities–Liquid or Pumpable Consistency (10^3 metric tons)

	1977	1983	1989
Organic Chemicals and Pesticides			
U.S., Gulf Coast	256	595	889
Canada	15	35	52
Mexico	8	31	86
Brazil	8	11	14
Japan	134	196	278
West. Europe	241	325	443
Subtotal	662	1193	1762
Inorganic Chemicals			
U.S., Gulf Coast	38	253	327
Canada	3	23	30
Mexico	1	1	3
Brazil	1	1	1
Japan	10	15	21
West. Europe	21	29	40
Subtotal	74	322	422
Petroleum Refining			
U.S., Gulf Coast	52	123	151
Canada	7	16	19
Mexico	3	8	20
Brazil	3	3	5
Japan	68	99	141
West. Europe	60	81	110
Subtotal	193	330	446
U.S. East Coast Waste	43	110	157
Eco. Total w/o U.S.	583	874	1263
Env. Total w/o U.S.[a]	547	805	1168
Eco. Grand Total	972	1955	2787
Env. Grand Total	896	1618	2346

[a]The environmental quantities do not include the inorganic chemicals.

ENVIRONMENTAL ACCEPTABILITY

The only toxic wastes that have been accepted for ocean incineration by IMCO and EPA are chlorinated liquid hydrocarbons. Complete combustion of such wastes yields carbon dioxide (CO_2), water (H_2O) and hydrogen chloride (HCl). Since complete combustion of unchlorinated hydrocarbons yields CO_2 and H_2O without HCl, it can be considered that such wastes are also implicitly approved.

Solid, containerized waste has not yet been approved for ship incineration. The attempt to incinerate such waste by the Matthias III in 1976 was unsuccessful. The 55-gal drums were consumed but the exhaust from the smokestack did not meet requirements. As far as is known, no further attempts at solid burns are now being planned.

Dissolved, slurried or suspended hydrocarbon wastes would essentially be considered as liquids and be acceptable for incineration. However, it would first be necessary to demonstrate that the incineration system involved met the combustion requirements. These include a 1250°C flame temperature, a minimum 1-sec residence time at 1250°C, a combustion efficiency (conversion of carbon to CO_2) of greater than 99.9% and a destruction efficiency of the toxic compounds of 99.95 ± 0.05%.

Aqueous or nonaqueous organic or inorganic salts would, in most instances, not be acceptable since they usually contain heavy metals such as lead, mercury, cadmium, arsenic and antimony. The London Dumping Convention, of which the U.S. is a signatory, forbids ocean disposal of heavy metals in any but trace amounts. The EPA has stated that it will also apply that requirement to incineration. This means that waste streams for ocean incineration are limited to part per million (ppm) concentrations of heavy metals. The precise value of the ppm concentrations of heavy metals has not been defined.

Toxic wastes which do not fall in any of the foregoing categories would have to be examined on an individual basis and the effect of their combustion products on the environment determined.

ECONOMICS OF OCEAN INCINERATION

The economics of incineration at sea will depend on the quantity of waste, the type of waste in terms of composition and physical form (liquid or solid), the type and capital cost of the incineration system to handle the required waste, the size and capital cost of the incineration vessel, and the operating costs. We have based the economic analysis on the economic waste quantities (i.e., waste streams from the organic chemical, pesticide, petroleum refining and inorganic chemical industries) and not on the environmental waste quantities (i.e., waste streams from the organic chemical, pesticide, and petroleum refining industries only). However, the impact on economic viability of deleting the inorganic waste streams and basing the analysis on the environmental quantities alone was also determined. Primary emphasis was on offshore incineration costs, although rough estimates were prepared for related onshore costs.

Study Assumptions

The basic assumptions that were used in performing the cost analysis are:

- Compliance with IMCO and EPA incineration ship requirements.
- Primarily nonaqueous, bulk, pumpable, chlorinated and unchlorinated organic

waste was considered. The average water content of the waste streams that were actually used in the cost calculations was 5%, and the average heating value was 6000 Btu/lb.

- Incineration ships are converted tankers with a toxic waste carrying capacity of 12,000 metric tons each and an average onsite incineration rate of 71 metric ton/hr after allowing for incineration downtime.
- The incineration ships will be constructed in the U.S. and will be U.S.-owned, -registered and -operated.
- Minimum two weeks per year in-port maintenance. For 1977 for the Gulf Coast the allowed in-port maintenance time period was 6.4 weeks based on economic waste quantities and 11.0 weeks based on environmental waste quantities.
- Unsophisticated incinerator operation and control based on prior EPA approved tests.
- Port turnaround time for waste pickup of 24 hr.
- Steaming speed of 15 knot to burn site; 2-knot speed while burning at burn site.
- Incineration ship operator owns and operates the incineration ships but arranges with tank truck, barge and tank farm operators for dedicated availability of their equipment and facilities.
- Fourth-quarter 1977 cost basis and uninflated dollars.
- Steady-state ship operation with overall yearly availability of 50% of the ship's incineration capacity. Start up and periodic EPA certification test costs were considered of minor impact and were not included in the ship incineration costs.
- A 15% return on capital employed and a 10% return on operating costs, after taxes. This resulted in an overall 41-52% after-tax return on invested capital.

Ocean Incineration Costs

For cost calculation purposes, a T2 tanker in the MarAd National Defense Reserve Fleet was conceptually converted to an incineration ship. The conversion was based on existing and planned USCG and IMCO regulations and codes for incineration ships and for ships carrying dangerous cargo in bulk. Acquistion, conversion and annual fixed and operating costs were determined, and the incineration cost per metric ton of waste was established for the U.S. Gulf and East Coasts for 1977, 1983 and 1989 based on uninflated 1977 dollars.

The capital cost for acquiring and converting a T2 tanker into an incineration ship, including the liquid incineration system, was estimated at $10.4 million. The yearly operating cost for the ship for 1977 in the Gulf Coast areas was $9.8 million. The operating cost was based on incineration in the EPA approved burn area 143 nautical miles southeast of Galveston and with waste pick-up at Mobile, New Orleans and Houston.

Incineration operations for 1983 in the Gulf Coast area are projected as requiring three 12,000-dwt vessels and include waste pick up at Veracruz, Mexico. This expands to four ships by 1989. Yearly operating costs for all of the ships for the three time periods, including a breakdown into the major cost categories, are shown in Table XXX.

Toxic waste quantities on the East Coast are seen as large enough by 1983 to permit initiation of operations by one 12,000-dwt vessel, which would also be sufficient for 1989. Pickup is postulated at Jersey City, Wilmington, Baltimore and Quebec, Canada.

Table XXX. Yearly Operating Cost, Gulf Coast (million $)

	1977	1983	1989
Crew Cost	4.3	12.8	17.0
Ship Fuel Cost	0.3	0.8	1.2
Incineration Fuel Cost	4.2	12.1	17.7
Insurance	0.3	0.9	1.2
Maintenance & Repair	0.7	2.2	2.9
TOTAL	9.8	28.8	40.0

Ocean incineration cost per ton of waste was calculated using:

$$C = \frac{A + B}{W} \tag{1}$$

where C = incineration cost per ton of waste
A = annual capital cost of the incineration ship(s)
B = annual operating cost
W = tons of waste incinerated per year

Annual capital cost was based on MarAd Title XI financing, 15% ROI, 10 year service life and straight-line depreciation. A 10% profit after taxes was imposed on the operating costs which means that a 1.1923 factor was applied to the operating costs to yield a 10% after-tax profit on operations. The incineration profit per metric ton of waste was $4-6, total profit increased from $1.4 million in 1977 to $6.9 million in 1989, and after-tax return on invested capital was 41-52%. The results of the calculations, including estimates of inland transportation (barge/truck) and storage costs, are shown in Table XXXI.

Sensitivity Analysis

A sensitivity analysis was performed on several potential cost variables and showed relatively little impact on incineration cost. For the Gulf Coast, the average increase in cost for incinerating a metric ton of waste is only 15% for each 100% increase in capital cost of the incineration vessel, 6% for each 100% increase in maintenance cost and 5% for each 100% increase in water content. There is essentially no change in cost from decreasing the quantity of waste by deleting the heavy metal-containing inorganic wastes from the total quantities.

Competition

Results from a study performed for the EPA under a separate contract [34] indicate that disposal costs by contractors for land incineration of chlorinated wastes range from $181 to $212/metric ton depending on Btu content. The study also indicates that the Vulcanus incineration ship is

Table XXXI. Cost and Profit Summary[a]

	Cost Elements[b] ($/metric ton)				After-Tax Profit	
	Inland Transp.	Storage	Lab. Anal.	Ocean Incin.	$/metric ton	Total ($)
				Gulf Coast		
1977	7.67	2.33	0.14	39.99	4.01	1,390,000
1983	7.67	1.12	0.05	40.46	4.07	4,100,000
1989	7.67	0.82	0.03	38.14	3.81	5,620,000
				East Coast		
1977						
1983		5.89	0.14	57.08	6.02	1,110,000
1989		4.19	0.10	50.29	5.09	1,320,000

[a]The costs in this table are based on the economic waste stream quantities. Corresponding ocean incineration cost for the Gulf Coast based on the environmental quantities (i.e., no inorganic waste streams) are $42.44/ton in 1977, $37.85 in 1983, and $37.56 in 1989.

[b]The cost elements have not been totaled so that comparisons can be made. For example, comparison of U.S. flag ship vs. foreign ship costs would be on the basis of the ocean incineration cost element only; comparison of ship incineration vs. land-based disposal would consist of the ocean incineration, storage and laboratory analysis costs vs. the land disposal contractor's charge and would not include transportation since transportation is required for both situations.

quoting $80-91/metric ton incineration charge for a minimum of two shiploads. There is no additional Europe-to-U.S. transportation charge for these quantities. This does not include storage or barge/truck charges for picking up the waste at the chemical plant and delivering it aboard ship. The $80-91 range is for normal, anticipated variation in Btu content of the waste. If the heat content is less than 5040 Btu/lb, the charge increases above $80/ton by the amount necessary to pay for the extra fuel to bring the heating value up to 5040 Btu/lb. There is also a 5-10% discount if waste quantity exceeds two shiploads. This would reduce the charge to $72-83/ton. The Vulcanus has a capacity approximately one-third that of a T2 tanker. It is interesting to note that the charge to the U.S. Air Force by the Vulcanus for disposing of the 10,400 metric tons of Herbicide Orange in 1977 was $250/ton including transit to and from Johnston Island in the Pacific. The charge for just the incineration portion of the operation, however, was $43/ton.

From the above, it appears the U.S. incineration ships would be competitive with European ships and with land incineration in the U.S., but that strong competition from foreign ships is a possibility.

CONCLUSIONS

The principal conclusions from the study are:

1. Ocean incineration of toxic chemical waste is economically and environmentally viable for U.S. flagships.

2. Incineration cost for U.S. flagships of 12,000 dwt waste capacity varies from $40 to $57/metric ton vs $80-91 for the Vulcanus and $181-212 for contracted, land-based incineration. Land-based incineration is more expensive because it requires scrubbers to remove the hydrochloric acid that is formed and caustic to neutralize the hydrochloric acid.

3. Based on the $40-57/metric ton cost, the after-tax profit to a U.S. incineration ship operator varies from $4 to $6/ton per ship and increases from $1.4 million in 1977 for one ship to $6.9 million in 1989 for five ships. This represents an after-tax return on investment of 41-52%.

4. The principal waste category for ocean incineration is represented by the chlorinated and unchlorinated liquid organic wastes. Solid waste has not been successfully incinerated at sea and has not been approved by EPA or IMCO.

5. Estimated net ocean incineration quantities of liquid and pumpable toxic waste available to U.S. ships are indicated below. These are the quantities for which ocean incineration is less expensive than alternative, acceptable, land-based disposal methods. Mexican, Canadian and U.S. East Coast wastes are included in the totals for 1983 and 1989:

346,000 metric tons in 1977
1,195,000 metric tons in 1983
1,733,000 metric tons in 1989

6. Net ocean incineration quantities of toxic waste available to U.S. ships which are environmentally acceptable as well as economically competitive are indicated below. These are wastes with only trace heavy metal content. Mexican, Canadian and U.S. East Coast wastes are included in the totals for 1983 and 1989:

308,000 metric tons in 1977
903,000 metric tons in 1983
1,355,000 metric tons in 1989

7. The number of incineration ships required to handle the toxic waste quantities (economic basis) increases from one in 1977, four in 1983, to five in 1989. These are vessels of T2 tanker size with a capacity of 12,000 metric tons of waste.

8. The principal location for generation of ocean incinerable waste is the Gulf Coast, which can support one ship now and four ships by 1989. The East Coast is next and can support one ship starting in 1983. The West Coast generates a negligible quantity of waste.

9. On an economic quantity basis (Item 5 above), the organic and pesticide industries produce 74% of the net ocean-incinerable waste, followed by the petroleum refining industry with 15% and the inorganic chemical industry with 11%. On an environmentally acceptable quantity basis (Item 6 above)

the organic and pesticide industries produce 83% of the waste and the petroleum refinery industry produces 17%.

10. Environmental requirements continue to become increasingly stringent. The major present impetus for ocean incineration is RCRA. RCRA will impose strict land disposal requirements for toxic wastes when it takes effect in July 1980. The estimated impact of RCRA was included in the waste quantity figures and numbers of ships noted in the foregoing.

11. International codes governing incineration ships and their design and operation were partially completed by IMCO in 1978. EPA and USCG will promulgate regulations for compliance with the codes.

12. There is minor impact on incineration cost from variation in ship capital cost, maintenance costs and water content of the waste, and in deletion of the inorganic chemical waste streams (i.e., heavy metals) from the incineration quantities.

REFERENCES

1. "Convention on the Prevention of Marine Pollution by Dumping of Wastes and Other Matter," (the London Convention), IMCO, completed 29 December 1972, entered into force 30 August 1975.
2. "Code for the Construction and Equipment of Ships Carrying Dangerous Chemicals in Bulk," IMCO, Resolution A.212 (VII), adopted 12 October 1971.
3. "Final Environmental Impact Statement, Maritime Administration Chemical Waste Incinerator Ship Project," MA-EIS-7302-76-041F, Vol. I and II, 2 July 1976.
4. Snider, W. D. "IMCO Conference on Tanker Safety and Pollution Prevention," *Maritime Reporter/Eng. News* (May 15, 1978).
5. "Tanker Safety and Pollution Prevention," *Federal Register*, 43(77 II) (1978).
6. "Safety Rules for Self-Propelled Vessels Carrying Hazardous Liquids," *Federal Register*, 42(186 III):49015–49046 (1977).
7. "Resource Conservation and Recovery Act of 1976," Public Law 94-580 [S.2150], October 21, 1976.
8. Claussen, E. L., S. A. Lingle and M. Newton. "Strategy for the Implementation of the Resource Conservation and Recovery Act of 1976," U.S. EPA, Office of Solid Waste, Draft Copy (1977).
9. "Waterfront Facilities," *Federal Register*, 43(69 V):15107–15116 (1978).
10. Arthur D. Little, Inc. "Physical, Chemical, and Biological Treatment Techniques for Industrial Wastes," Report SW-148c, Contract No. 68-01-3554, U.S. EPA, Office of Solid Waste Management Programs (1976).
11. TRW. "Assessment of Industrial Hazardous Waste Practices: Organic Chemicals, Pesticides and Explosives Industries, Final Report," SW-118c, Contract No. 68-01-2919, U.S. EPA, Office of Solid Waste Management Programs (1976).
12. Processes Research, Inc. "Alternatives for Hazardous Waste Management in the Organic Chemical, Pesticides and Explosives Industries, Final

Report," Contract No. 68-01-4127, U.S. EPA, Office of Solid Waste Management Programs (1977).
13. Versar, Inc. "Assessment of Industrial Hazardous Waste Practices, Inorganic Chemicals Industry, Final Report," SW-104c, Contract No. 68-01-2246, U.S. EPA, Office of Solid Waste Management Programs (1975).
14. Versar, Inc. "Alternatives for Hazardous Waste Management in the Inorganic Chemicals Industry, Final Report," Contract No. 68-01-4190, U.S. EPA, Office of Solid Waste (1977).
15. Calspan Corp. "Assessment of Industrial Hazardous Waste Practices in the Metal Smelting and Refining Industry," Calspan Report No. ND-5520-M-1, Contract No. 68-01-2604, U.S. EPA, Office of Solid Waste (1977).
16. Calspan Corp. "Alternatives for Hazardous Waste Management in the Metals Smelting and Refining Industries, Final Report," Contract No. 68-01-4312, U.S. EPA, Office of Solid Waste (1977).
17. Jacobs Engineering Co. "Assessment of Hazardous Waste Practices in the Petroleum Refining Industry, Final Report," Contract No. 68-01-2288, U.S. EPA, Office of Solid Waste Management Programs (1976).
18. Jacobs Engineering Co. "Alternatives for Hazardous Waste Management in the Petroleum Refining Industry, Final Report," Contract No. 68-01-4167, U.S. EPA, Office of Solid Waste (1978).
19. *Business Statistics 1975* U.S. Department of Commerce, Bureau of Economic Analysis (May 1976).
20. "Survey of Current Business," (monthly update to *Business Statistics*), U.S. Department of Commerce, Bureau of Economic Affairs.
21. "U.S. Industrial Outlook," U.S. Department of Commerce, Domestic and International Business Administration (January 1977).
22. "Clean Water Act of 1977," Public Law 95-217, 27 December 1977.
23. "Clean Air Act Amendments of 1970," Public Law 91-604, 31 December 1970, amended August 1977.
24. "National Air Quality and Emissions Trends Report, 1976," Docu. No. EPA-450/1-77-002, U.S. EPA, Office of Air and Waste Management (1977).
25. "Tough Cleanup Rules Set for Chemical Dumps," *Chem. Eng.* (18 December 1978), pp. 56-58.
26. *Chem. Eng. News* (21 November 1977).
27. *Oil Gas J.* (21 November 1977).
28. *Chem. Eng.* (5 December 1977).
29. *Oil Gas J.* (26 December 1977).
30. *Oil Gas J.* (21 November 1977).
31. "Synthetic Organic Chemicals, United States Production and Sales, 1973," United States International Trade Commission, ITC Publication 728 (1975).
32. "Synthetic Organic Chemicals, United States Production and Sales, 1976," United States International Trade Commission (1977).
33. "World Chemical Outlook '78," *Chem. Eng. News* (19 December 1977), pp. 30-55.
34. Shih, C. C. et al. "Comparative Cost Analysis and Environmental Assessment for Disposal of Organochlorine Wastes," U.S. EPA Contract No. 6A-02-2613, EPA-600/2-78-190 (1978).

CHAPTER 14

DISPOSAL OF HAZARDOUS WASTES BY MOLTEN SALT COMBUSTION

S. J. Yosim, K. M. Barclay, R. L. Gay and L. F. Grantham
Rockwell International
Energy Systems Group
Canoga Park, California

INTRODUCTION

Our complex industrial society generates millions of tons of various types of industrial waste materials every year. A sizable fraction of this waste is considered hazardous. Previous methods for disposal or treatment of these hazardous wastes for disposal have included landfill, ocean dumping, deep-well disposal and conventional incineration, but some of these methods are undesirable because they pollute the air, water or land. Thus, alternative methods to these traditional means of disposal are being investigated.

This chapter presents some experimental results which demonstrate the feasibility of applying molten salt combustion technology to the disposal of different types of hazardous wastes. The concept of molten salt combustion is described first, followed by a description of the molten salt combustors at Rockwell International (RI). Some results of molten salt combustion tests on a variety of hazardous wastes are given. A brief discussion of disposal of the used salt is then presented. A summary of the advantages of molten salt combustion concludes the chapter.

CONCEPT OF MOLTEN SALT COMBUSTION

In the RI concept for molten salt combustion (Figure 1), combustible material and air are continuously introduced beneath the surface of molten sodium carbonate at 800-1000°C. The combustible material is added in such a manner that any gas formed during combustion is forced to pass through the

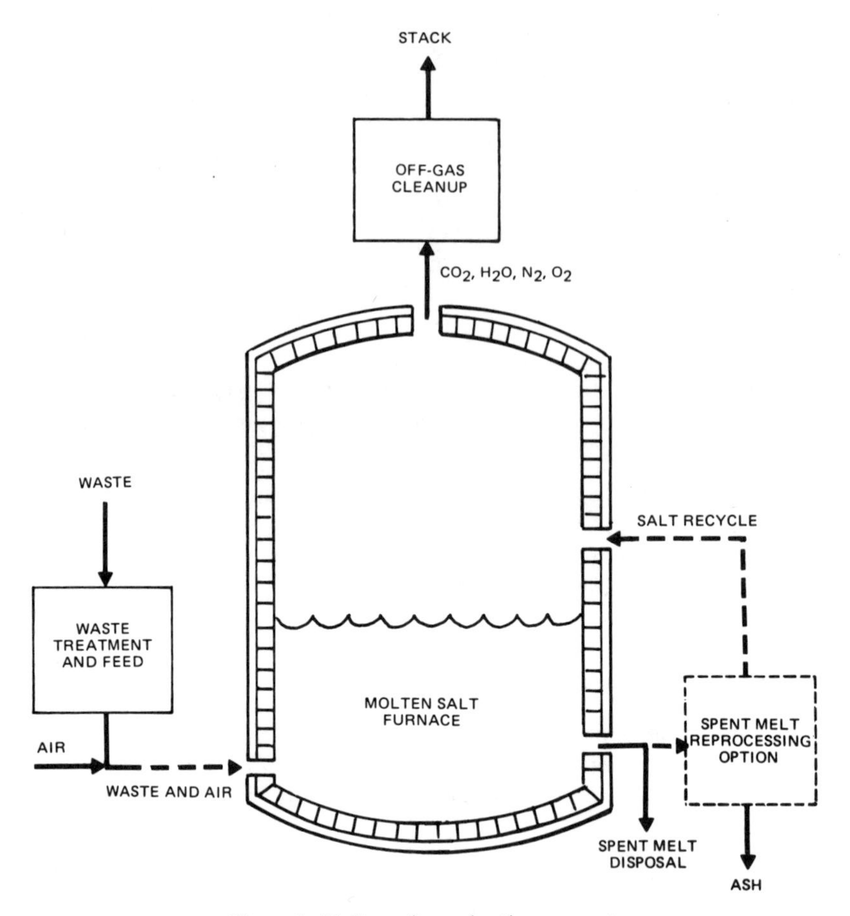

Figure 1. Molten salt combustion concept.

melt before it is emitted into the atmosphere. The offgas containing carbon dioxide, steam, nitrogen and unreacted oxygen is cleaned of particulates by scrubbing in a venturi scrubber or by passing it through a baghouse.

Inorganic products resulting from the reaction of organic halogens, phosphorus, arsenic, etc., with sodium carbonate buildup in the melt and must be removed. This is to prevent complete conversion of the sodium carbonate with an eventual loss of the acid pollutant removal capability. Ash, introduced by the waste, must be removed to preserve the fluidity of the melt. An ash concentration of about 20 wt% provides an ample margin of safety for the melt fluidity. In certain applications with low throughput, the melt is removed in batches. When the throughput is sufficiently large, a side-stream of the melt is withdrawn either in batches or continuously and is processed. The melt can be quenched in water. The solution is then filtered to remove the ash and processed to recover the carbonate which is then recycled to the combustor.

Sodium carbonate is used because it is compatible with combustion products CO_2 and H_2O and reacts with acidic gases such as HCl (produced from organic chloride compounds) and SO_2 (from organic sulfur compounds). It is stable, nonvolatile, inexpensive and nontoxic. The chemical reactions of the waste with salt and air depend on the waste composition. The carbon and hydrogen of the waste are converted to CO_2 and steam. Halogens form their corresponding sodium halide salts. Phosphorus, sulfur, arsenic and silicon (from glass or ash in the waste) form the oxygenated salts, Na_3PO_4, Na_2SO_4, $NaAsO_3$ and Na_2SiO_3, respectively. The iron from metal containers forms iron oxide. The temperatures of combustion are too low to permit a significant amount of nitrogen oxides to be formed by fixation of the nitrogen in the air. The ash is trapped in the melt.

MOLTEN SALT COMBUSTORS

There are two molten salt combustion facilities used at RI for tests with waste. One is a bench-scale molten salt combustor with feed rate of 0.25-1 kg/hr of waste. Feasibility and optimization tests are usually carried out in this combustor, which was used for most of the tests reported in this paper. The other is a pilot-plant combustor, capable of processing 25-100 kg/hr of waste, and is used to obtain engineering data for the design of an actual plant. (Another existing molten salt combustor is the DOE-funded coal gasification process development unit which has a design throughput of 1000 kg/hr of coal.)

Bench-Scale Combustor

A schematic of the bench-scale combustion system is shown in Figure 2. The details of the system are described elsewhere [1-3]. Basically, 5.5 kg of molten salt are contained in a 15-cm i.d., 90-cm high alumina tube placed in a Type 321 stainless steel retainer vessel. This stainless steel vessel, in turn, is contained in a 20-cm i.d., 4-heating-zone Marshall furnace. Details of methods of feeding dry solids, combustible liquids, aqueous solutions, slurries and wet wastes are presented elsewhere [1-3]. Briefly, solids are pulverized and metered with a screw feeder, liquids and aqueous wastes are pumped with liquid pumps and sprayed, and slurries are fed with a slurry pump.

Pilot-Scale Molten Salt Combustor

A photograph of the molten salt pilot combustor is shown in Figure 3. The design details are given elsewhere [1,2]. Briefly, the molten salt vessel is 4 m high, 0.85 m i.d. and is made of Type 304 stainless steel lined with

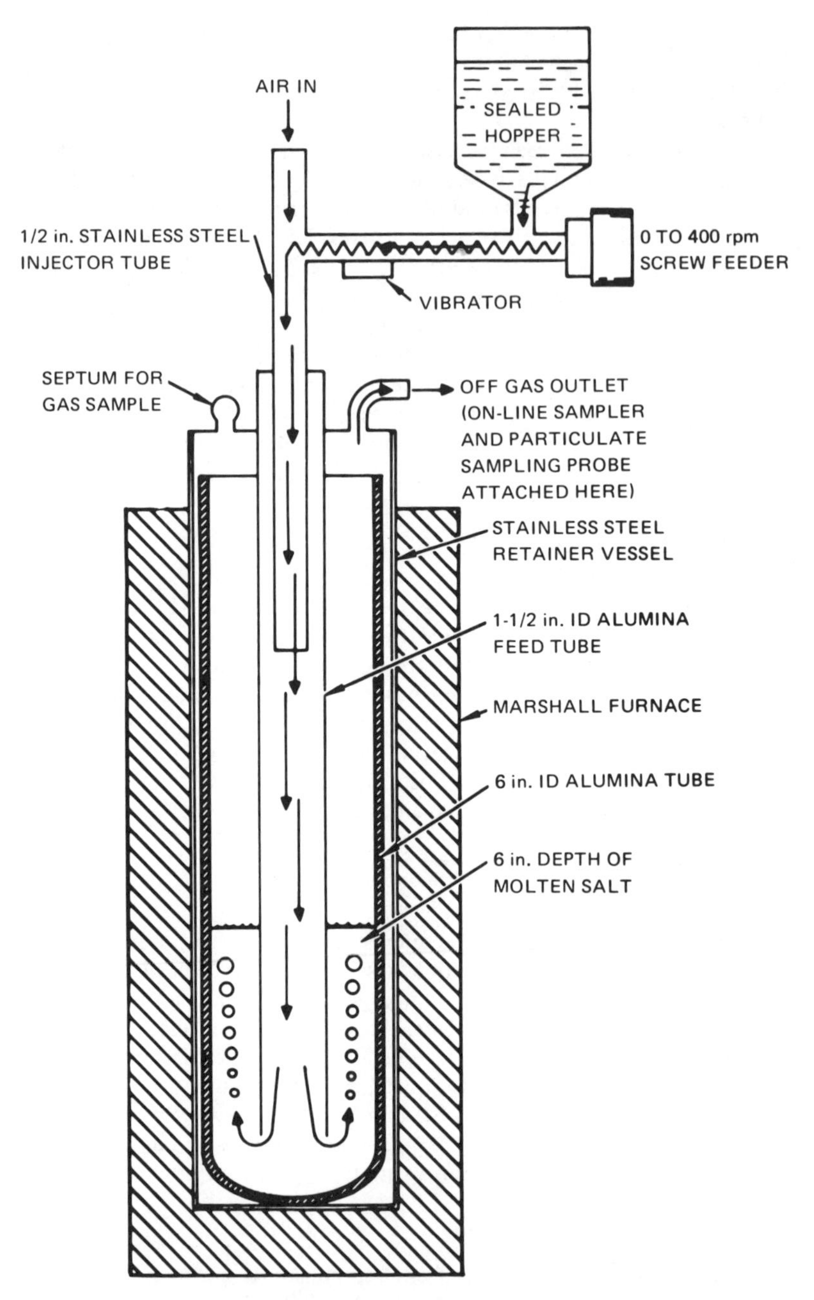

Figure 2. Bench-scale molten salt combustor.

Figure 3. Pilot-scale molten salt combustor.

15-cm-thick refractory blocks. It contains 1000 kg of salt, which corresponds to a quiescent melt depth of 0.80 m. The vessel is preheated on startup and kept hot on standby by a natural gas-fired burner. The heat content of the waste is, in general, sufficient to maintain the salt in the molten state during actual combustion.

RESULTS OF COMBUSTION TESTS

A list of the hazardous materials that have been disposed of by molten salt combustion is shown in Table I.

Chemical Warfare Agents

A series of molten salt combustion tests was performed on undiluted agents at Edgewood Arsenal by their personnel [4]. The bench-scale molten salt combustor was designed and built by RI. The tests were performed at 900-1000°C on VX ($C_{11}H_{26}O_2PSN$), GB ($C_4H_{10}O_2PF$) and mustard ($C_4H_8Cl_2S$). The results are shown in Table II. No agents were detected in the melt, particulate filter or stack gas. Thus, the concentration values for the agents in the stack gas and on the filter represent upper limits. (The relatively high upper limit for VX on the filter is because the filter residue was analyzed by a gas chromatography/mass spectrometry technique which is far less sensitive than the enzymatic procedure used for the stack gases.) A comparison of the results with the stack emission guidelines proposed by the Army shows that the stack emissions for GB and mustard in general meet these guidelines. If a more sensitive analytical procedure for VX determination had been used, it might have been possible to demonstrate that the stack emission guidelines for this agent also would have been met. The extent of destruction of these agents is shown in the last column of Table II. These values were based on the assumption that any agent not found in the stack or volatiles was destroyed by the salt. The extent of destruction was greater than 99.99998%, 99.99999% and 99.9999997% for mustard, VX and GB, respectively.

Carbon monoxide, unburned hydrocarbons and NO_X emissions were, in general, less than 0.1%, 10 ppm and 50 ppm, respectively. No SO_2 emissions were detected for VX and mustard (<0.1 ppm). In the case of GB, the NaF

Table I. Hazardous Materials Disposed of by Molten Salt Combustion

Chemical Warfare Agents	Pesticides
GB	DDT
VX	Malathion
Mustard	2,4 D
Chemical Wastes	Combustible Waste from Nuclear Plant
PCB	Tributyl Phosphate
Perchloroethylene Bottoms	Ion Exchange Resins
Trichloroethane	Silicon Carbide
Chloroform	Leaded Gloves
Diphenylamine·HCl	Aqueous Solutions
Nitroethane	
Paraarsanilic Acid	

Table II. Destruction of Chemical Warfare Agents by Molten Salt Combustion[a]

Agent	Compound Emission (ppt)		Compound Emission (mg/m^3)		Emission Guidelines Stack + Filter (mg/m^3)	Extent of Destruction of Agent (%)
	Stack	Filter	Stack	Filter		
VX	<1.2	<470	<0.000012	<0.0049	0.00003	>99.9999945
	<5.8	<1140	<0.000060	<0.012	0.00003	>99.999989
GB	<27.5	<45.6	<0.00012	<0.00023	0.0003	>99.99999969
	<11.5	<33.7	<0.00005	<0.00017	0.0003	>99.99999976
Mustard	<3460	<39.3	<0.023[b]	<0.00026	0.03	>99.999985
	<3160	<36.3	<0.021[b]	<0.00024	0.03	>99.999982

[a]Tested at Edgewood Arsenal by their personnel.
[b]The relatively low sensitivity in the stack gas from mustard was due to the fact that in these tests the offgas was split into three scrubbers to analyze for HCl and SO_2 emissions as well.

and HF emissions were about 20 and 1 ppm, respectively; the P_2O_5 emissions were, in general, below 1000 ppm. The particulate levels were about 0.2, 0.6 and 3.0 g/m^3 for VX, mustard and GB, respectively.

Edgewood Arsenal personnel also performed tests on the agent mustard mixed with dunnage material such as metal, plastic, glass and wood. These tests were performed to test the feasibility of destroying toxic gas identification sets used by the armed services in training for the recognition and detection of various war gases. In these tests, the salt initially contained 20 wt% NaCl and 10% ash (Fe_2O_3 and SiO_2) as well as Na_2CO_3 to simulate steady-state conditions. To prepare feedable material, the liquid mustard was adsorbed on activated carbon, which was then mixed with the appropriate amounts of dunnage. Again, no agents were detected. This test also demonstrates that activated carbon used to adsorb very toxic materials can be effectively disposed of by molten salt combustion.

Pesticides [1]

Tests were carried out at about 900°C on DDT powder ($C_{14}H_9Cl_5$), malathion ($C_{10}H_{19}O_6PS_2$) dissolved in xylene and 2,4-D (an ester of dichlorophenoxyacetic acid). The melts contained either Na_2CO_3 or K_2CO_3. The use of K_2CO_3 is of interest because the excess K_2CO_3 and the reaction products KCl, K_3PO_4, and K_2SO_4 can be used as fertilizer. Typical combustion results on DDT and malathion are shown in Table III. Destruction of the pesticide was greater than 99.99%. No pesticides were detected in the melt; however, traces of pesticides were sometimes detected in the offgas. The last two columns of Table III compare the concentration of pesticide in the offgas with threshold limit values (TLV). (TLV refers to airborne concentrations of

Table III. Typical Results of Combustion Tests on Malathion and DDT

Pesticide	Salt	Pesticide Destroyed (%)	Concentration of Pesticide in Melt (ppm)	Quantity in Exhaust Gas (mg/m^3)	TLV[a] of Pesticide (mg/m^3)
DDT	Na_2CO_3	99.998	ND[b] <0.05	0.3	1
DDT	K_2CO_3	99.998	ND <0.2	0.3	1
Malathion	Na_2CO_3	99.9998	ND <0.01	0.06	15
Malathion	K_2CO_3	99.999	ND <0.005	ND <0.4	15

[a]Threshold limit value.
[b]Not detected.

substances and represents conditions under which it is believed that nearly all workers may be exposed, day after day, without adverse effect.) The concentrations of pesticides in the exhaust gas were generally well below the TLV. This comparison is a conservative one since the gas will be considerably diluted when it reaches the worker area. Another consideration is the fact that in these tests, a 15-cm deep salt bed was used. In an actual disposal plant, a 1-m deep salt bed is expected to be used. This will increase the residence and contact time; therefore, the extent of destruction of these pesticides is expected to exceed considerably the 99.99+% found in the laboratory tests.

It is not clear why the extent of destruction in the case of DDT and malathion was not as good as in the case of the chemical warfare agents. It may be that the pesticides are more refractory than the agents. However, it is interesting to note that polychlorinated biphenyls (PCB), which are also known to be refractory, were also completely destroyed by molten salt combustion (see below).

The herbicide 2,4-D was of interest because it was an actual waste which contained 30 to 50% 2,4-D, and 50 to 70% tars (mostly *bis*-ester and dichlorophenol tars). The waste, which was rather viscous, was diluted with ethanol to reduce the viscosity. The destruction of the herbicide at 830°C was >99.98%; no organic chlorides or HCl were detected in the melt or in the exhaust gas. (In this case, a less sensitive analytical technique was used than for DDT.)

Combustion tests have also been performed on materials which could be used for pesticide containers. The materials tested were paper, plastic (an equal-weight mixture of polyvinyl chloride (PVC) and polyethylene) and rubber. Destruction of the combustible material was rapid and complete [1]. The CO ($\leqslant$0.2%), NO_x (<65 ppm) and unburned hydrocarbons (<30 ppm) in the offgas were very low. No HCl was detected in the tests with PVC.

Combustion tests on 700 kg of material, much of it similar to that of pesticide containers, were performed in the pilot combustor at feed rates of about 30 kg/hr [6]. The waste contained paper (53 wt%), polyethylene (32%) PVC (8%) and rubber (7%). No HCl (<5 ppm), SO_2 (<2 ppm), CO (<0.1%) or hydrocarbons (<0.1%) were detected in the offgas. The NO_x concentration was about 30 ppm. The O_2, CO_2 and N_2 concentrations were 5-12%, 10-15% and 76-78%, respectively.

Chemical Wastes

Halogenated Chemicals

Tests have been performed on PCB, perchloroethylene-containing waste slurries, trichlorethane and chloroform.

PCB. PCB are environmentally stable compounds which are considered to be carcinogenic and teratogenic. In general, special incinerators operating at high temperatures (1200°C) and long residence times (2 sec) are required to destroy these materials. Tests were performed to determine if PCB could be destroyed by molten salt combustion. The PCB tested had a composition corresponding to the empirical formula $C_{9.08}H_{4.00}Cl_{4.00}$. Tests were performed at different temperatures (700-980°C), air/PCB ratios (90-230% of stoichiometric air) and melt composition. Melts which contained various amounts of NaCl and K_2CO_3 in addition to the Na_2CO_3 were used. (The K_2CO_3 was added to lower the melting point of the salt; the NaCl was formed in situ by reaction of the chlorine of the waste with the carbonate.) No PCB (<50 $\mu g/m^3$) were detected in any of the offgas samples, even when all the carbonate had been converted to sodium chloride; thus, the extent of destruction exceeded 99.9999%. As long as some carbonate was present (>3 wt% Na_2CO_3), no organic material was detected in the offgas. Since the melt height was 6 in. and the superficial gas velocity was 1 ft/sec, the nominal residence time of the PCB in the salt was 0.5 sec. Thus, the results indicate that molten salt combustion is a very effective means of destroying PCB completely and rapidly even when the temperature is as low as 700°C and the residence time is as short as 0.5 sec.

Perchloroethylene-Containing Wastes. * Perchloroethylene (C_2Cl_4) is a common industrial solvent and is used in the nuclear industry. During the refabrication of high-temperature, gas-cooled reactor fuel, perchloroethylene is used as the scrubbing medium to remove condensible hydrocarbons, carbon soot and uranium-bearing particulates from the offgas streams. Eventually the perchloroethylene becomes degraded and must be purified by distillation. The bottoms from this distillation is a waste stream. Since this waste stream contains a large number of complex polynuclear aromatic hydrocarbons (PAH), some of which are known carcinogens, it is necessary to completely destroy this material. A program was carried out to test the feasibility of destroying such a waste by molten salt combustion; the waste contained about 93 wt% perchloroethylene, 6% organic degradation products and about

*This work was performed for Oak Ridge National Laboratory, Order No. 25X-49341V.

1% solids (carbon, silicon carbide, etc.). The slurry was fed with a peristaltic pump. Due to the low heating value of the waste (less than 7000 j/g), kerosene was added to the waste to furnish auxiliary heat to maintain melt temperature. Parameters such as temperature (850-950°C), stoichiometry (30-100% excess air) and perchloroethylene-kerosene weight ratio (7-9) were varied. No peaks of organic material above background were found in any sample. If any organic compounds in the offgas came from the perchloroethylene bottoms, their concentrations were, in general, less than 0.5 mg/m^3. Therefore, within the limits of the chemical analyses of these tests, destruction of perchloroethylene bottoms by molten salt combustion was complete.

The concentration of HCl in the offgas was less than 2 ppm as long as any Na_2CO_3 (~1%) remained in the melt. The concentration of NO_x in the offgas was also very low, less than 25 ppm. The particulate content of the exhaust gas increased with time and with increasing temperature, which was expected since the particulates consisted mainly of NaCl and were due to vaporization of NaCl from the melt.

Trichloroethane and Chloroform. A summary of the results from the combustion of chloroform and trichloroethane is shown in Table IV. No unreacted materials were found in the offgas of those tests. The extent of destruction exceeded 99.999%. In the case of trichloroethane, prevention of HCl emissions was accomplished with as little as 2 wt% Na_2CO_3 in the melt.

A series of tests was conducted with chloroform in the molten salt pilot plant. However, instead of feeding the chloroform continuously, bulk quantities of chloroform contained in metal canisters were plunged into the melt. The apparent advantage of such a feed technique is the elimination of comparatively heavy, bulky, noisy and costly shredding equipment when

Table IV. Summary of Results of Combustion Tests of Chemical Wastes

Chemical	Average Test Temperature (°C)	Chemical Destroyed (%)	Concentration of Chemical in Melt (ppm)	Quantity in Gas (mg/m^3)	TLV (mg/m^3)
Chloroform	818	>99.999	<0.1	<0.5	50
Trichloroethane	840	>99.999	ND[a]	<1.7	1900
Diphenylamine·HCl	922	>99.999	<0.1	<0.4	20
Nitroethane	892	>99.993	<1	<4.4	310
Paraarsanilic Acid	924	>99.999	<0.1[b]	<0.8[c]	0.5

[a]Not determined.
[b]While the concentration of paraarsanilic acid in the melt is extremely low, the concentration of inorganic arsenic compounds in the melt is, as expected, high, since the arsenic is retained in the melt as sodium arsenate.
[c]Reported as inorganic arsenic.

dealing with sealed bottles or cans of chemicals. Attendant maintenance and decontamination considerations are also important. In these tests, 3-4 sealed Pyrex vials containing a total of 160 cm^3 of chloroform were packed in cardboard tubes or sawdust and placed in 10-cm diameter metal canisters which were sealed. Thus, they simulated war gas training kits. The canister was plunged beneath the surface of the melt. About 30 combustion tests were performed. As long as the metal canister was punctured to allow melt access to the glass vials containing the chloroform, little, if any, noise was heard; noise was heard only with nonpunctured canisters. With nonpunctured canisters, the vials ruptured after about 1-3 min in the melt, whereas when the canisters were punctured, the vials ruptured less than 1 min after the canister was immersed in the melt. Drager tubes were used to monitor the offgas; no hydrogen chloride (<1 ppm), phosgene (<1 ppm) or chloroform (<5 ppm) were detected in the offgas. Small amounts of chloroform were detected in all benzene scrubber samples analyzed; however, the amount of chloroform found was near the detection limits. The extent of destruction of chloroform in these pilot plant tests was, in general, >99.95% with both punctured and unpunctured canisters.

Nitrogen-Containing Chemicals

Nitrogen-containing compounds during combustion can emit large concentrations of NO_X in the offgas due to the nitrogen in the feed. Tests were performed on diphenylamine·HCl ($C_{12}H_{12}NCl$), monoethanolamine (C_2H_7ON) and nitroethane ($C_2H_5NO_2$) to determine the extent of destruction and/or to determine the conditions for minimizing NO_X release. A summary of the results of the extent of destruction of diphenylamine HCl and nitroethane is shown in Table IV; no unreacted material was detected. Since these nitrogen-containing compounds did produce substantial amounts of NO_X formed from the nitrogen in the feed, tests were performed on monoethanolamine and nitroethane to minimize NO_X formation. It was found that by suitably adjusting the air/waste feed ratio, the NO_X concentration was sharply reduced. To illustrate, the NO_X concentration in the offgas from the combustion of monoethanolamine decreased from 2200 ppm at 265% theoretical air (165% excess air) to 200 ppm at 108% theoretical air, to 20 ppm at 75% theoretical air. (The 75% of stoichiometric air constitutes combustion in a semireducing condition.)

In the case of nitroethane, the extent of NO_X formation was greater than in the case of the monoethanolamine, with the NO_X concentration = 16,000 ppm at 107% of stoichiometric air. However, with more reducing conditions, than with the amines, the NO_X concentration could be lowered to a value less than 100 ppm.

Organic Arsenic Compound

Tests have been performed on paraarsanilic acid ($C_6H_8NAsO_3$). A summary of the results on paraarsanilic acid is included in Table IV. No paraarsanilic acid was detected in the melt or in the offgas. Although no paraarsanilic acid was detected in the melt, the reaction product sodium arsenate, as expected, was retained in the melt. Thus, it is recognized that in this case, the arsenic-containing melt must be considered to be hazardous. One approach for the disposal of this particular melt would be to solidify it into an unleachable glass. (Solidification of the spent salt into an unleachable form is described below.)

Combustible Waste From Nuclear Plants

Combustion tests were performed on materials from nuclear plants which are, in general, difficult to burn. These include tributyl phosphate ($C_{12}H_{27}PO_4$), ion-exchange resins, silicon carbide (SiC), leaded gloves and aqueous solutions containing $NaNO_3$ and NaOH.

Tributyl Phosphate

Tributyl phosphate (TBP) and its radiolysis products dibutyl and monobutyl phosphate are waste products resulting from Purex processing of nuclear fuels. Both pure TBP and diluted TBP (30% TBP in kerosene) were combusted at 900°C with 45 and 28% excess air, respectively. Both feeds were consumed rapidly, as evidenced by the high CO_2 and very low CO and hydrocarbon concentrations in the offgas. The CO_2 content of the offgas was 10-14%; the CO, unburned hydrocarbons and NO_x concentrations were 0.05%, 20 ppm and 30 ppm, respectively.

Ion-Exchange Resins

Ion-exchange resins are also difficult to burn in a conventional incinerator; they generally produce a smoky flame containing evolved hydrocarbons. Tests were performed with Dowex-1 and Powdex; both were of the styrene divinyl benzene cross-linked polymer type containing the trimethyl amine (anion) or sulfonic acid (cation) grouping as the active component. Again, the materials burned rapidly and completely.

Silicon Carbide

Silicon carbide hulls are produced in reprocessing high-temperature gas-cooled reactor fuels. This material is difficult to dispose of by regular incineration because a protective film of SiO_2 builds up on the particles and inhibits combustion. However, SiO_2 is soluble in molten Na_2CO_3, and thus, the protective film can be destroyed by the melt. Particles of SiC 50 μ in diam were burned in the melt. At 900-960°C, SiC was converted to SiO_2. At 1000°C and above, the SiC was consumed rapidly and completely by formation of Na_2SiO_3. Apparently, the conversion of SiC to SiO_2 is more rapid than the conversion of SiO_2 to Na_2SiO_3 in the 900-960°C range. It was concluded that SiC can be effectively disposed of in molten Na_2CO_3 provided the temperature is 1000°C or higher.

Leaded Gloves

Leaded gloves from a glove box are also difficult to burn in a conventional incinerator; further, a great deal of lead is emitted as a gas. Two molten salt combustion tests at 880°C were performed with leaded gloves containing 55 wt% Pb in the form of PbO. Combustion was rapid and complete; the CO, hydrocarbon, NO_x and SO_2 (from the sulfur in the gloves) concentrations were 0.05%, 10 ppm, 38 ppm, 20 ppm and not detected, respectively.

After each test, a pool of molten elemental lead was found on the bottom of the combustor. Less than 0.5% of the lead in the feed was found in the offgas. In an actual system, the lead in the offgas could be removed by a baghouse.

*Aqueous Solutions**

During the reprocessing of nuclear fuels by the Purex process, a considerable amount of intermediate-level aqueous waste is formed. This waste generally consists of concentrated general purpose aqueous waste from various sources, spent solvent from Purex extraction, discarded ion exchange resin and miscellaneous solid waste (such as PVC) from decontamination and maintenance. A program was performed to determine the technical feasibility of disposing of such intermediate-level liquid radioactive waste by molten salt combustion. A simulated (nonradioactive) waste consisting of 96.6 wt% aqueous solution (4 *M* $NaNO_3$ and 4 *M* NaOH) 1.0% PVC, 1.6% 30% TBP in kerosene and 0.8% ion-exchange resin was combusted. Sufficient methanol (to form a 28.7 wt% solution) was added to supply enough heat to maintain the melt temperature and to vaporize the water during the combustion. The CO_2 from the methanol also served to convert the NaOH and $NaNO_3$ to Na_2CO_3. The test of the simulated intermediate-level waste slurry containing 4 *M* NaOH and 4 *M* $NaNO_3$ was successfully performed with a water to methanol weight ratio of 1.5. The concentrations of CO and unburned hydrocarbons in the offgas were less than 200 ppm. The NO and NO_2 concentrations were 6400 ppm and 2000 ppm, respectively. In an actual plant, the offgas would be combined with the effluent stream of the Purex-nitric acid dissolving stream. Therefore, reduction of the NO_x levels was not of primary concern. Complete destruction (99.99%) of the organics in the slurry mixture was obtained.

Radioactive Wastes

Tests have been performed on wastes contaminated with radioactivity to destroy completely the organic fraction and to reduce the volume of the waste. The radioactive wastes that have been treated are transuranic wastes, fission product-contaminated wastes and activation product-contaminated wastes. The results of such tests have been described elsewhere [5,6];

*This work was performed for Oak Ridge National Laboratory, Order No. 89X-22353V.

DISPOSAL OF SALT AFTER COMBUSTION

The spent salt after combustion consists essentially of soluble alkali salts, e.g., NaCl, Na_2SO_4 and Na_3PO_4, resulting from reaction of the Cl, S and P of the waste with Na_2CO_3, as well as unreacted Na_2CO_3. Local conditions may require insolubilization of the salt. Therefore, tests to insolubilize the salt in glass and concrete were performed.

Salt samples were glassified using different fluxing material as the solidifying agent. The salt (Na_2SO_4 and NaCl) content of the solidified glass was varied from 10-40 wt%. Glassification temperatures varied from 1100-1500°C. The glassified samples had leach rates of 10^{-4}-10^{-6} g/cm^2/day in an accelerated leach test (continually-flowing distilled water at 100°C); this is equivalent to 10^{-7}-10^{-9} g/cm^2/day in normal leach tests. It would appear that clay, Pyrex and probably basalt can be used to satisfactorily glassify waste salts from molten salt combustion of organic wastes.

Waste salts containing Na_2CO_3, Na_2SO_4 and NaCl were mixed with cement to test insolubilization in concrete. The salt content of the salt-cement mixture was 10-38 wt%. The leach rates of the concrete samples were only about 5-10 times greater than those of glassified waste salt. Based on these results, it would appear that cement formation (which is less expensive than glassification) would be an acceptable method of disposing of waste salts.

CONCLUSIONS

Molten salt combustion has been shown to be applicable to a wide variety of hazardous wastes. These wastes include chemical warfare agents, hazardous industrial wastes, pesticides and carcinogenic materials. The process is applicable to several waste forms, i.e., solids, liquids, solid-liquid mixtures and gases. The process is attractive for hazardous wastes because of its capability for complete and rapid destruction. This rapid destruction results from the catalytic effect of the salt and the contact of the air and waste with the hot molten salt, which provides rapid transfer of heat to the waste. Other features of the molten salt process are (1) sodium carbonate, the main constituent of the melt, is stable, nonvolatile, inexpensive and nontoxic; and (2) no acidic gaseous pollutants, e.g., HCl from chlorinated compounds and SO_2 from sulfur-containing compounds, are emitted. This is true even when only small amounts of sodium carbonate (~1 wt%) remain in the melt.

However, it should be pointed out that there are some wastes for which this process is not particularly suited—from an economic viewpoint, rather than from a technical viewpoint. Examples of such wastes are: (1) wastes with a high ash content (e.g., >20% ash), since ash leads to increased melt viscosity which results in more frequent changing of the salt and, therefore, increased salt makeup; and (2) municipal wastes and sludges which are not particularly hazardous and for which less expensive processes are probably available.

The molten carbonate is known to be corrosive. However, dense alumina has been found to be an excellent container material for molten salt combustion. The corrosion-resistance of alumina is due to the formation of sodium aluminate which is only slightly soluble in sodium carbonate (~25 ppm at 1000°C) and which forms a self-healing protective film.

Molten salt combustion has the advantage over conventional incinerators in that scrubbers to remove corrosive acidic gases from the offgas are not necessary. The advantage over fluidized beds containing solids capable of reacting with acidic gases is that the degree of use of the melt is very high while that of the solid particles used in the fluidized bed is considerably lower since, in the latter case, a coating of neutralization product impedes further reaction. The very high degree of destruction of chemical warfare agents suggests that molten salt combustion is the most effective method for destroying hazardous wastes and is probably the most suitable method when complete destruction is of paramount importance.

REFERENCES

1. Yosin, S. J., K. M. Barclay and L. F. Grantham. "Destruction of Pesticides and Pesticide Containers by Molten Salt Combustion," in *Disposal and Decontamination of Pesticides*, M. V. Kennedy, Ed. (Washington, DC: American Chemical Society, 1978), p. 118.
2. Yosim, S. J. and K. M. Barclay. "Production of Low-Btu Gas from Wastes Using Molten Salts," in *Fuels From Waste*, L. L. Anderson and D. A. Tillman, Eds. (New York: Academic Press, Inc., 1977), p. 41.
3. Yosim, S. J., K. M. Barclay, R. L. Gay and L. F. Grantham. "Disposal of Laboratory Hazardous Wastes by Molten Salt Combusion," in *Safe Handling of Chemical Carcinogens, Mutagens, Teratogens and Highly Toxic Substances*, D. B. Walters, Ed. (Ann Arbor, MI: Ann Arbor Science Publishers, Inc., 1980).
4. Dustin, D. F., M. Riordan, E. Vigus and A. Wagner. "Applications of Molten Salt Incineration to the Demilitarization and Disposal of Chemical Material," EM-TR-76099 (1977).
5. Grantham, L. F., and D. E. McKenzie. "Molten Salt Combustion of Radioactive Wastes," in *Proceedings of the Fourteenth ERDA Air Cleaning Conference*, CONF-760822, p. 17 (1976).
6. Grantham, L. F., D. E. McKenzie, R. D. Oldenkamp and W. L. Richards. "Disposal of Transuranic Solid Waste Using Atomics International's Molten Salt Combustion Process II," AI-ERDA-13169, NTIS, Springfield, VA (1975).

CHAPTER 15

COMPOSTING OF INDUSTRIAL WASTES

E. Epstein and J. E. Alpert
Energy Resources Company Inc.
Cambridge, Massachusetts

INTRODUCTION

Composting is the rapid biological degradation of organic materials under controlled conditions. All organic compounds are potentially biodegradable. The rate and extent of biodegradation varies, however, with the chemical nature of the compounds and the conditions under which the compound is degraded. The biodegradation of industrial compounds by soil organisms is well documented. Kaufman [1] and Guenzi [2] studied the biodegradation of pesticides. Kaufman [1] showed that position, types and number of halogen substitutes are important factors affecting the rate of decomposition of aliphatic and some aromatic pesticides. Murthy et al. [3] and Murthy and Kaufman [4] showed that phenols are biologically degraded in soils. Bushnell and Haas [5] found that *Pseudomonas* spp. was one of the predominant metabolizers of crude oil. Considerable literature exists on the microbial degradation and use of crude oils by soil organisms [6,7]. Generally, the decomposition of organic compounds in soils is slow.

Landfarming involves the incorporation of the organic material into the top horizons of the soil where microbial degradation takes place. This activity often occurs at suboptimal conditions of moisture, temperature, oxygen, nutrients or other parameters. Composting, on the other hand, is the controlled biodegradation of organic material. Although composting can occur over a wide range of conditions, most composting is more efficient in an aerobic mode. The moisture content should range from 40 to 60% on a wet-weight basis. At moisture contents below 40% the process slows down, whereas above 60% the free pore space is reduced and improper aeration occurs. The ideal carbon-to-nitrogen ratio is between 20 and 30, with oxygen levels between 5 and 15%. Carbon-to-nitrogen ratios in excess of 30 will result in a slower process, whereas below 20, nitrogen losses are high. Oxygen levels below 5% can result in anaerobic pockets and excessive oxygen (above 15%) will affect the temperature regime. Composting can occur under mesophilic or thermophilic conditions. The microflora vary under these two conditions. This variation can be used to advantage in attempts to decompose different compounds as will be indicated later.

Composting may represent one potential way of accelerating industrial organic waste decomposition.

COMPOSTING OF INDUSTRIAL MATERIAL

Historically, composting has been used to biodegrade and stabilize animal, human or municipal solid waste. Recently [8] there has been considerable interest in a system which rapidly composts sewage sludge under aerobic thermophilic conditions. Presently there are very few data on the composting of industrial organic compounds. Rose and Mercer [9] investigated the composting of insecticides in agricultural wastes. Figure 1 shows the data for the pesticides diazinon and parathion (phospharothioate insecticides). Continuous thermophilic composting rapidly decreased their concentration. In 10 days the diazinon concentration was reduced approximately 50% and in 42

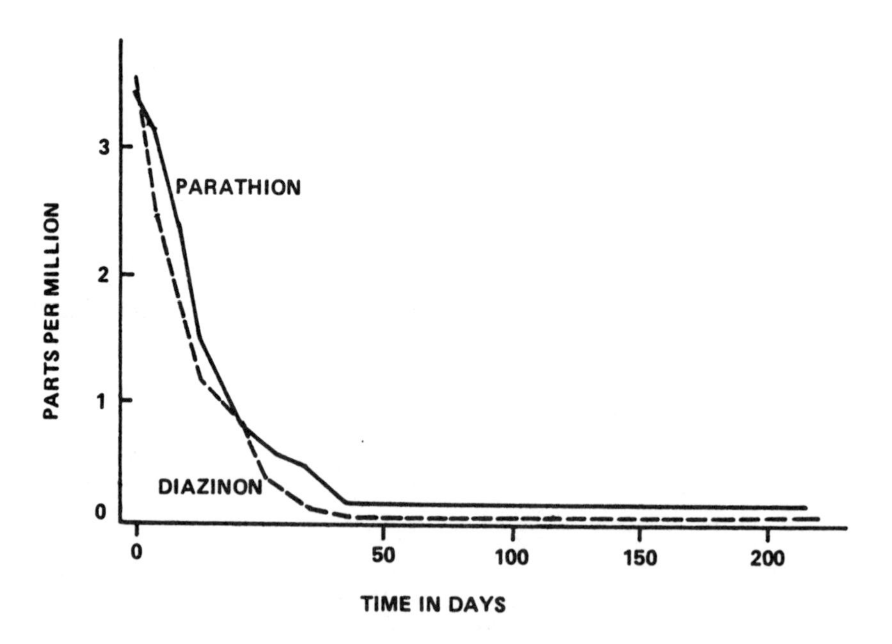

Figure 1. Reduction in the concentration of the pesticides diazinon and parathion during composting.

days the initial concentration of 3.3 ppm was reduced to less than 0.002 ppm. The concentration of parathion was reduced by 50% in 12 days of composting. Degradation of pp-DDT (Figure 2) was considerably slower. Fifty days of composting were required to reduce the level of DDT from 2.2 to 0.8 ppm. It was interesting to note that DDD and DDE, which are normally produced by DDT degradation in soils, were not detected in the composted material. This aspect is extremely important since breakdown products such as DDD and DDE may be as environmentally undesirable as the parent compound.

Osmon and Andrews [10] investigated the biodegradation of TNT, a nitro-amino-toluene compound, by composting. They were able to degrade large quantities (5-10%) of TNT in a relatively short time to zero or acceptable levels (Figure 3). No breakdown products or compounds of environmental concern were found.

As part of a sewage sludge composting project for the Boston Metropolitan District Commission (MDC), Energy Resources Co. Inc. (ERCO) has been investigating the effect of composting on the degradation of hydrocarbons in sewage sludge. Sludge extracts were fractionated into classes of compounds and a detailed analysis was performed on the aromatic fraction. The aromatic fraction was chosen for analysis because it would contain polynuclear aromatic hydrocarbons (PAH), many of which are known or suspected carcinogens.

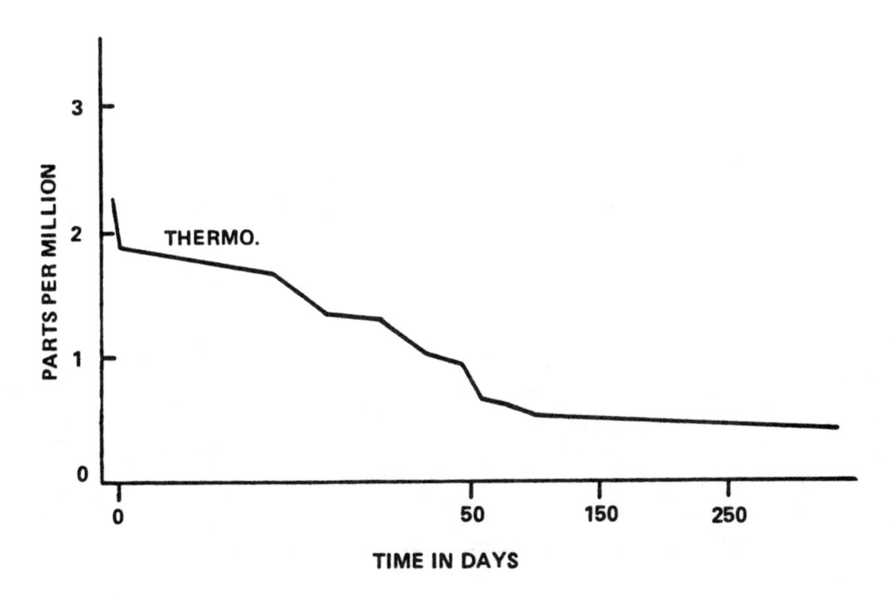

Figure 2. Reduction in the concentration of pp-DDT during composting.

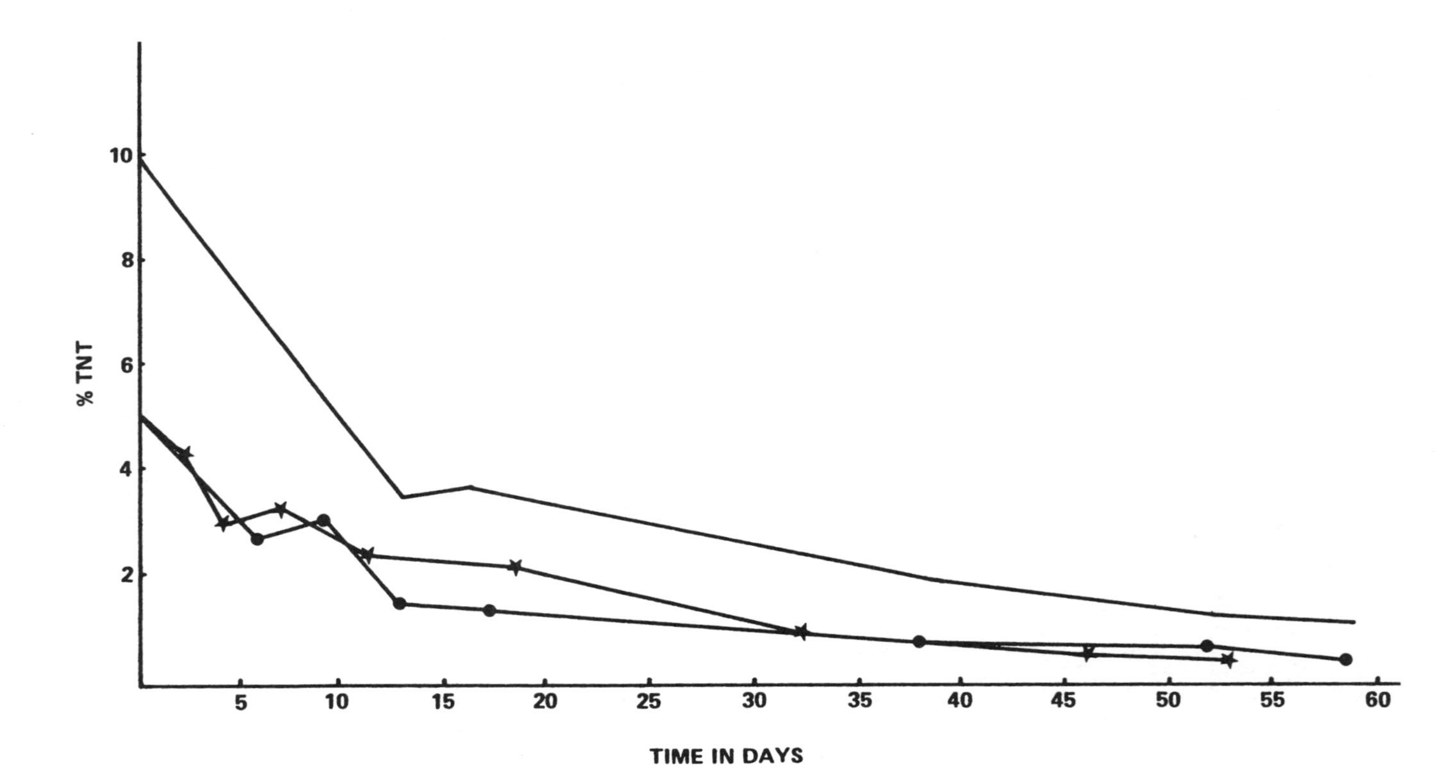

Figure 3. Degradation of TNT during composting. The effect of different initial TNT loadings on biogradation rate.

Figure 4 shows chromatograms for raw and composted sludge from MDC's Deer Island treatment plant. Comparison of the two chromatograms shows that composting almost completely degraded the hydrocarbons. The single large peak in the composted sludge chromatogram has not yet been identified. Data from gas chromatography and mass spectrometry show that composting at Deer Island removed all but one of the resolved aromatic compounds, all of the resolved aliphatic components and a sizable portion of the unresolved components from the raw sludge.

This analysis represents only the aromatic fraction, one of the many classes of organic compounds contained in sludge. Further work is necessary to ascertain whether other classes of compounds could be biodegraded

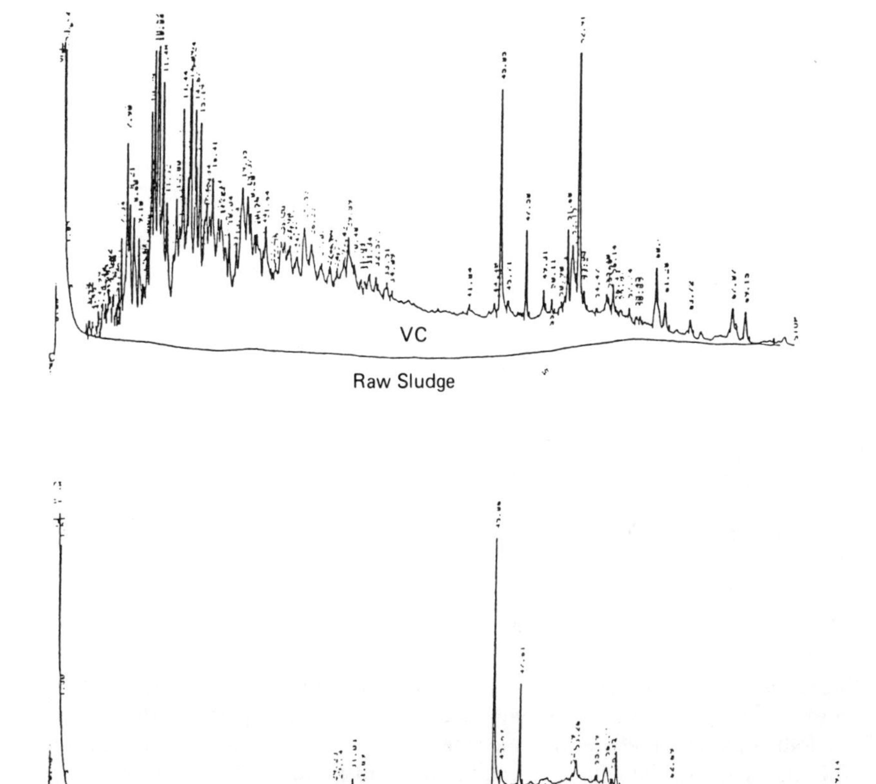

Figure 4. Gas chromatographic analysis of organic compounds present in raw sewage sludge before and after composting.

through composting. Data on the aromatic fraction indicate the potential of composting not only for stabilizing putrescible material but also for degrading or decomposing environmentally hazardous organics.

There are numerous nonhazardous industrial wastes which are produced in large volumes. Recently, ERCO investigated the composting of pulp- and papermill sludges [11]. The sludges contained 45-55% inorganic matter with a large proportion of lime. In many cases these sludges contain very low levels of heavy metals thus making the final product suitable for agriculture and horticulture. The presence of chlorinated hydrocarbons which may pose environmental problems may be reduced or eliminated by composting. Currently under a U.S. Coast Guard (USCG) grant ERCO is investigating the composting of oily wastes.

The following is a partial list of some of the more potentially biodegradable industrial wastes.

1. organic chemicals and solvents
    - phenol
    - xylol
    - toluene
    - ketones, alcohols
2. oils, fats and waxes
    - bunker oil
    - petroleum acid pitch
    - waste oil and diesel fuel
3. wood and paper products
    - papermill sludge
    - paper wastes
    - wood wastes–bark, shavings
4. agricultural and pharmaceutical wastes
    - molasses silage
    - mycellial wastes
    - apple, grape and other fruit pomace
    - food wastes

COMPOSTING TECHNIQUES

Accelerated aerobic composting can be achieved by three methods: (1) windrow, (2) enclosed systems, and (3) forced aeration.

The windrow method has been used predominantly by agriculture for stabilizing animal wastes or in preparation of bedding material for the mushroom industry. A windrow turning machine aerates the organic matter by periodic mixing. Temperatures are generally lower, with wide variation among different parts of the windrow, which can result in nonuniform biodegradation of the organic matter. If pathogens are present, destruction may be incomplete. Enclosed systems have been used in Europe for composting of animal and solid wastes. The high capital costs as well as operational problems has not resulted in their use in the United States. In 1975, the U.S. Department of Agriculture (USDA) developed the forced aeration system for composting sewage sludge [8]. The major advantages of this method for handling wastes are:

1. The microbial decomposition of sludge during composting alleviates malodors and produces a stable, humus-like, organic material.
2. Heat produced during composting effectively destroys human pathogens.
3. The product can be stored conveniently, and spread easily and uniformly on land. Sludge, on the other hand, is difficult to handle and produces odors when stored.
4. Composting produces an organic resource that can be used beneficially on land as a source of plant micro- and macronutrients, and as a soil conditioner.
5. Low capital investment and low energy requirements.

Figure 5 shows a flow diagram for the Beltsville aerated pile method. Filter cake sewage sludge (approximately 22% solids) is mixed with a bulking material and then composted in a stationary aerated pile for 21 days. Bulking materials which have been used include: woodchips, shredded paper, paper briquets, leaves, peanut hulls and automobile salvage fabrics. The paper and paper briquets were obtained from a resource recovery plant after air classification. Automobile salvage fabric wastes consisted of shredded foam rubber and fabric material recovered after metal removal.

Sludge at this moisture content (78% water) will not compost aerobically alone because sufficient air cannot penetrate the biomass, either by diffusion or forced aeration. The bulking material provides the necessary texture, structure and porosity for aeration lowers the moisture content of the biomass to about 50 or 60% and provides an additional carbon source for microorganisms to ensure rapid composting. The ratio of the bulking material to sludge varies with the type of material and its physical characteristics, and the moisture content of the sludge. For composting sludge with woodchips, two volumes of woodchips are mixed with one volume of sludge (about 1:1 on a weight basis). This produces the necessary absorbency to lower the moisture content from 78% to about 60%. Mixing can be achieved with various types of equipment, the simplest being a front-end loader.

After mixing, the aerated pile is constructed (Figure 6). A loop of 10-cm (diam) perforated plastic pipe is laid on the surface and covered with 30 cm of woodchips or unscreened compost. This is the base of the pile. The sludge-woodchip mixture is then placed in a pile on the prepared base, after which the pile is blanketed with a 30-cm layer of screened compost for insulation and odor control. Fifty tons of wet sludge produces a pile with a triangular cross section, and approximate dimensions of 16 m long, 7 m wide and 2.5 m high. The loop of perforated pipe is then connected with solid pipe to a 0.33-hp blower controlled by a timer. Aerobic composting conditions are maintained by drawing air through the pile. The effluent airstream is conducted into a small pile of screened, cured compost, where odorous gases are effectively absrobed. The blower is operated intermittently, to provide oxygen levels between 5 and 15% during the composting period. The sludge-woodchip mixture is composted in the pile for 21 days, during which time the sludge is

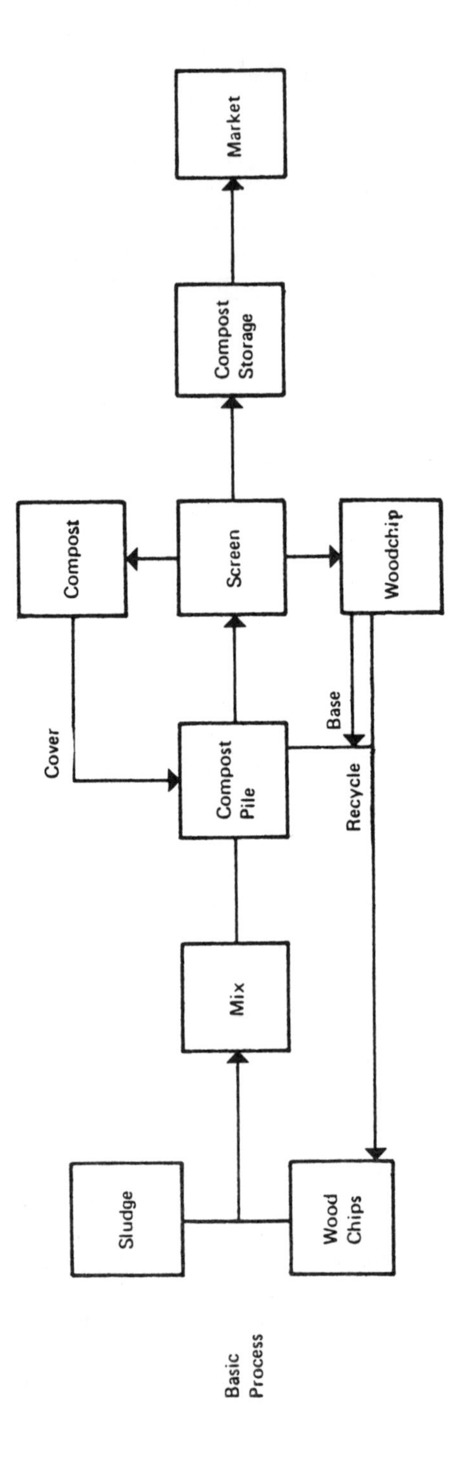

Figure 5. Composting process flow.

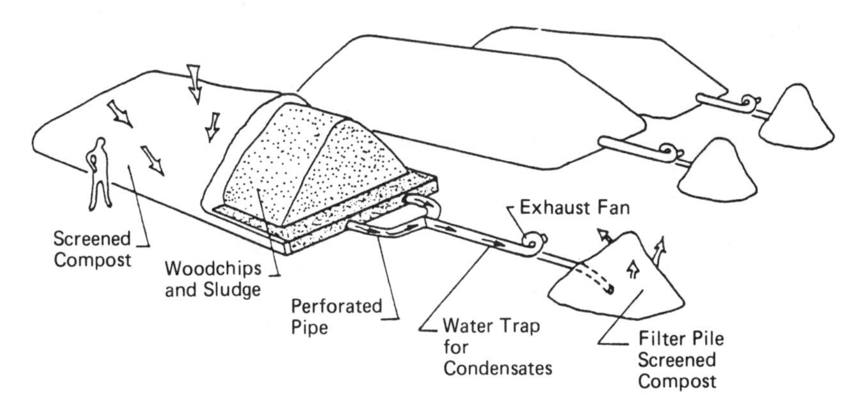

Figure 6. Three-dimensional schematic diagram of the Beltsville aerated pile method for composting sewage sludge.

stabilized by the rapid decomposition of volatile organic solids, and odors are abated.

A modification of the aerated pile is the aerated extended pile in which each day's sludge delivery is mixed with woodchips and placed in a pile which uses the shoulder (lengthwise dimension) of the previous day's pile, forming a continuous or extended pile. The extended pile reduces the operating area by 50%. Moreover, the amount of blanket material required for insulation and odor control is also decreased by 50%, as is the woodchip requirement for the pile base.

After the composting period and depending on the composition of the waste it may be necessary to further stabilize it by additional curing. Curing is an extension of the composting process, although temperatures are generally lower. Drying and screening are often carried out to remove and recycle the bulking material. Screening may also be desired to produce a marketable product.

CONCLUSION

The disposal of industrial wastes by conventional systems such as landfilling or incineration is becoming less attractive to industries. This is the result of legislative actions requiring more stringent control of wastes to prevent environmental impacts and the increased cost of fossil fuel used for incineration and waste transportation.

Biological degradation of organic industrial wastes through composting may result in efficient, economic and safe disposal of the wastes. There is a need to evaluate the parameters which would accelerate the biodegradation of various materials.

REFERENCES

1. Kaufman, D. D. "Structure of Pesticides and Decomposition by Soil Microorganisms," in *Pesticides and Their Effects on Soils and Waters*, ASA Special Pub. No. 8. (Madison WI: Soil Science Society of America Inc., 1966).
2. Guenzi, W. D. *Pesticides in Soil and Water* (Madison, WI: Soil Science Society America Inc., 1974).
3. Murthy, N. B. K., D. D. Kaufman and G. F. Fries. "Anaerobic and Aerobic Soil Degradation of Pentachlorophenol," *J. Environ. Sci. Health*. B14(1): 1-214 (1979).
4. Murthy, N. B. K., and D. D. Kaufman. "Degradation of Pentachloronitrobenzene (PCNB) in Anaerobic Soils," *J. Agric. Food Chem.* 26:1151-1156 (1978).
5. Bushnell, L. E., and H. F. Haas. "The Utilization of Certain Hydrocarbons by Microorganisms," *J. Bact.* 41:1082-1089 (1941).
6. Jobson, A. M., F. D. Cook and D. W. S. Westlake. "Microbial Utilization of Crude Oil," *J. Appl. Microbiol.* 23:1082-1089 (1972).
7. Kincannon, B. C. "Oily Waste Disposal by Soil Cultivation Process," U.S. EPA, Washington, DC, EPA R2-72-110 (1972).
8. Epstein, E., G. B. Willson and J. F. Parr. "The Beltsville Aerated Pile Method for Composting Sewage Sludge," in *New Processes of Wastewater Treatment and Recovery*, G. Mattock, Ed. (England: Society of Chemical Industry, Ellis Harwood Ltd. Pub.), 1978.
9. Rose, W. W., and W. A. Mercer. "Fate of Insecticides in Composted Agricultural Wastes," Progress Rept. Part I. National Canners' Association, Washington, DC (1968).
10. Osmon, J. L., and C. C. Andrews. "The Biodegradation of TNT in Enhanced Soil and Compost Systems," Tech. Rept. ARLCD-TR-77032. U.S. Army Armament Res. and Dev. Command, Dover, NJ (1978).
11. Epstein, E. "Composting Pulp and Paper Mill Wastes," paper presented at NCASI Northeast Regional Meeting, November 1-2, 1978, Boston MA.

CHAPTER 16

SEABED DISPOSAL OF RADIOACTIVE WASTE–A PRIME ALTERNATIVE

N. Sonenshein
Global Marine Development Inc.
Newport Beach, California

INTRODUCTION

Radioactive wastes have been commonly categorized as high-level (HLW), intermediate or transuranic (TRU), and low-level wastes (LLW). The quantities on hand are significant, and are expected to increase steadily with growing numbers of commercial nuclear power plants. Recent developments in deep-sea technology such as deep-seabed drilling and emplacement, and recovery of large items of equipment make it possible to consider disposing of nuclear wastes in deep ocean sediments. Seabed disposal for HLW is based on a multibarrier concept that would contain it long enough for it to decay to background levels. It is focused on the consolidated, fine-grained sea sediments located in Atlantic and Pacific midplate, midgyre (MPG) areas. Disposal of LLW is a simpler problem because shielding and thermal dissipation are not necessary. The potential advantages for both HLW and LLW include remoteness from human activity, isolation capabilities of the ocean sediments, high heat-sink capacity of the oceans, the large areas available and avoidance of the problems of gaining access to adequate land sites.

This thesis will be developed in this chapter by discussion of the following topics:

- classification of radioactive wastes
- quantities of radioactive wastes
- disposal techniques
- deep-sea technology
- seabed disposal of HLW/spent fuel
- seabed disposal of LLW/mill tailings

CLASSIFICATION OF RADIOACTIVE WASTES

One of the most elusive aspects of discussing radioactive waste management is classifying the wastes for which disposal is contemplated. Categorizations in use are based on waste composition, disposal method and source of the wastes.

By Composition [1]

One perspective, for purposes of definition, can be gained by noting that radioactive wastes are generated by the process within a fission reactor. Two classes of chemical elements comprise such wastes:

1. Fission products are fragments produced when a heavy nucleus is split, and are responsible for the fairly rapid generation of heat during the early years of the waste lifetime; and
2. Actinide elements consist of uranium which has not undergone fission, transuranic elements formed by neutron capture and the decay products of these two types of elements. Actinides generally have longer half-lives than fission products and generate considerably less heat/mass/time.

By Disposal Method [2]

A new approach at classification that is now under study by the U.S. Nuclear Regulatory Commission (NRC) is based on the disposal method that has been selected as most appropriate for the waste being considered. This leads to three methods for handling wastes:

1. Discharge directly to the biosphere in a manner similar to the handling of routine trash;
2. Confine the waste for a period of time in a controlled manner with predictably low release rates; and
3. Isolate the waste from the biosphere so that biologically significant releases or inadvertent reentry by mankind into the disposal area is highly unlikely. One can conclude that wastes appropriate for interim confinement can be taken as LLW, and that wastes that must be isolated from the biosphere are considered HLW.

By Source

Most commonly, wastes are classified by source, as HLW, TRU and LLW on the basis of their penetrating radiation. High-level solid wastes require shielding to limit exposures of management personnel within acceptable limits and also require means or removal of decay heat during storage to avoid large thermal effects on the storage medium. Intermediate-level solid wastes require shielding for personnel protection but do not need heat removal for safe storage. Low-level wastes can be handled and stored in their original waste containers without added shielding.

HLW, as stated in the foregoing, is derived from the aqueous product of the first stage of fuel reprocessing, but the current U.S. policy of not reprocessing as a deterrent to nuclear weapon proliferation has impaired this definition in the U.S. for the present. Thus, one must consider irradiated reactor fuel elements–spent fuel–together with already existing HLW.

Mill tailings generated by uranium mining constitute another source of waste that is characterized by large volumes and low radioactivity.

In this chapter radioactive wastes will be classified by the common source definitions, LLW, TRU, HLW and mill tailings, as defined above.

QUANTITIES OF RADIOACTIVE WASTE [3]

The quantities of radioactive waste accumulated to date and expected to be generated by the year 2000, including commercial- and defense-generated wastes, are summarized in the following. Where data or conversion factors can be conveniently applied, the data are presented in cubic feet of waste material. It will be noted, however, that such normalization is not practical in every instance, and the reader's indulgence is requested for information presented in "mixed" units. Two cases are presented to provide a range of possible quantities for nuclear capacities projected for the year 2000: Case 1–148 GWe; Case 2–GWe.

LLW Quantities

These typically consist of filters, used equipment, ion-exchange resins, clothing, trash, rubble and the like.

- Existing–76% Dept. of Energy (DOE): 66.6 million ft^3
- Cumulative, Year 2000–Case 1: 121 million ft^3; Case 2: 450 million ft^3

TRU Quantities

These wastes all have greater radioactivity than 10 nCi/g, and are buried or stored. If buried, they are mixed with sand, gravel, soil, contaminated metal, concrete and parts.

- Existing–Commercial: 123 kg (buried, volume not known); DOE: 1100 kg
- Cumulative, Year 2000–Case 1: 6.8 million ft^3; Case 2: 116 million ft^3

HLW/Spent Fuel

The high-level waste is alkaline or acid, and is liquid or solid crystallized salt or sludge. It is a dry powder if calcined, and is leachable in water or acids. "Spent fuel" is irradiated fuel elements that have not been reprocessed.

1. HLW:
    - Existing (99% DOE)–9.48 million ft^3
    - Cumulative, Year 1985–9.1 million ft^3
2. Spent Fuel:
    - Existing–0.03 million ft^3
    - Cumulative, Year 2000, including 10% of foreign–Case 1: 1.22 million ft^3; Case 2: 1.57 million ft^3

Mill Tailings

These wastes are generated in large quantities as a result of uranium mining, and are now stored aboveground.

- Existing 140 million tons
- Cumulative, Year 2000–Case 1: 370 million tons; Case 2: 485 million tons

Recapitulation of Waste Quantities

	Existing	Case 1 Yr 2000	Case 2 Yr 2000
LLW, million ft^3	66.6	121	450
TRU, million ft^3	1223 (kg)	6.8	116
HLW, million ft^3– Yr 1985	9.48	9.1	9.1
Spent Fuel, million ft^3	0.03	1.22	1.57
Mill Tailings, million tons	140	370	485

One would like to aggregate these data further, but the lack of uniformity of units makes this impossible. Nevertheless, one can appreciate that the quantities involved in value are now significant and are projected to increase substantially.

DISPOSAL TECHNIQUES

Conventional

LLW and TRU

LLW and TRU wastes are currently disposed of in the U.S. by shallow land burial. This method has been used for other than HLW since the inception of nuclear weapons research in the U.S. in the 1940s. "Burial" refers to the placement of waste at relatively shallow depths in earth materials without provision for later retrievability. In 1946, the U.S. started ocean dumping of LLW, with most of the activity occurring between 1946 and 1962. Growing pressures against this latter technique coupled with the availability of alternative land sites (for a total of five DOE and six commercial) led, in 1971, to the discontinuance of ocean dumping by the U.S. Burial in land sites now faces severe restrictions because of evidence of radionuclide dispersion into the biosphere through groundwater migration. Only two commercial sites (Barnwell, SC and Beatty, NV) are now accepting wastes, and the disposal of LLW has become a critical issue in the operation of commercial nuclear power plants. A Congressional report [4] issued by the House Committee on Government Operations found that "the issue of storage and disposal of these low level wastes remains as one of the major unresolved issues of today, and it will assume even greater proportions as the quantity of such waste material continues to grow."

HLW/Spent Fuel

No permanent disposal of these wastes has yet been effected. Previously reprocessed waste (military and commercial) is in storage, and spent fuel elements are stored in water pools near reactor sites awaiting disposal. The opinion of the informed scientific and technical community [3] is that a practical solution can be achieved. The solution requires the use of mined repositories in a deep geologic, hydrogeologic, geochemical and tectonic environment such as rock salt, shale, granite and the like. The DOE program for radioactive waste management now places prime reliance on this approach; however, public acceptance has been negative–11 states have already imposed legal barriers to such installations within their boundaries.

Nonconventional

Seabed

The seabed approach

> . . . is focussed on the unconsolidated, fine-grained, deep-sea sediments as candidate disposal media for containment of radionuclides from solidified high level radioactive wastes. These sediments, which are generally described as "red clays", possess several characteristics that make them attractive: (1) the sorption properties tend to inhibit movement of radionuclides; (2) the low permeability of these clays cause the rate of water migration (natural or induced) to be very low; (3) the strength of these materials is relatively low, thus facilitating subsurface emplacement of waste containers; and (4) the sediments which are in a plastic state and do not fracture if disturbed and tend to "heel" or flow if disrupted [5].

This disposal concept contemplates implanting canisters of waste several tens of meters into such deep-ocean sediments by means of a free fall penetrometer, or by other means such as deep-sea drilling and emplacement.

Very Deep Drill Holes

This method is similar in approach to mined repositories. This concept would use very deep drill holes for emplacing canisters of HLW/spent fuel in a column about 4000 ft long that is at least 25,000 feet below the earth's surface. Conceptual design and safety analyses have not yet been carried out, but much of the same work as is needed for mined repositories is applicable to this approach, which depends on large extensions of current technology [3].

Rock Melting

This approach contemplates disposing of aqueous solutions and slurries of solidified waste from reprocessing plants by pumping them through a drilled hole into a cavity in a very deep and relatively impermeable geologic formation. There they would dissolve and become part of the formation. Suspension of reprocessing in the U.S. and lack of feasibility detract at this time from acceptance of this approach [3].

Outer Space

Space disposal has been studied extensively because it offers the prospect of putting the material into the sun, where extreme temperatures would provide for ultimate disposal free of future human concern. Deterrents to this approach are its very high cost and the danger of launch accidents that could result in worldwide dispersal of radioactive materials.

Transmutation

This concept would partition transuranic isotopes from fission products by recycling in a nuclear reactor. This necessarily involves spent fuel reprocessing and plutonium recycling. It is possible that the result of such partitioning and transmutation would not appreciably alter the overall risk to society from the disposal operation because some long-lived fission products from the transmutation would have to be separated and disposed of geologically. This approach is some 10-15 years from achievement [3].

Appraisal

The relative attractiveness of each of the nonconventional concepts was realistically appraised by the President's Interagency Preview Group [3] when it stated:

> With respect to R & D on technical options other than mined repositories, the nearer term approaches (i.e., deep ocean sediments and very deep holes) should be given funding support so they may be adequately evaluated as potential competitors. Funding for rock melting, space disposal and transmutation would allow some feasibility and design work to proceed.

When one considers that radioactive waste management is now the most critical factor in the nuclear fuel cycle, the wisdom of concentrating its development program at one site and in one geologic medium is questionable. High priority should certainly be accorded to the preferred site and the method that has been furthest developed (mined geologic repositories); but alternative sites and concepts need to be evaluated concurrently. Conversely, the number of alternatives should be limited to a manageable few, among which seabed disposal is a prime contender.

DEEP-SEA TECHNOLOGY

Existence of deep-sea technology, such as deep-seabed drilling, drill hole reentry, deep-sea mining and deep-sea emplacement and recovery of large equipment, makes it possible to consider disposing of nuclear wastes within deep-ocean sediments. Development of this technology in the last 20 years has been rapid, with the pace accelerating as the offshore search has intensified for hydrocarbons and minerals to augment the world's shrinking supplies. Two ships–Glomar Challenger and Glomar Explorer–exemplify the most advanced aspects of ship design and performance in deep-sea engineering, and will be described briefly to illuminate the state-of-the-art in this rapidly moving technology.

Glomar Challenger (Figure 1)

This dynamically positioned drilling ship, that takes its name from a historic British pioneer in oceanographic research, is 400 feet long and displaces some 10,500 tons. Delivered in 1968 to Global Marine Inc. of Los Angeles, CA, who designed, supervised construction of in Orange, TX, and has been operating Challenger for the National Science Foundation (NSF) this ship is the principal scientific tool of the international Deep Sea Drilling Project.

Figure 1. Glomar Challenger.

Under the management of Scripps Institution of Oceanography, Glomar Challenger:

> "has sailed 250,000 nautical miles and has drilled in all the world's oceans including the Antarctic, the Mediterranean, and the Black Sea. To 1 July 1977, 429 sites were occupied and drilled; 625,000 feet of ocean bottom have been drilled, and 159,000 feet of core recovered and archived. The operation record has been spectacular: the vessel had only 100 down days in 10 years, not counting the planned changeover and overhaul periods" [6].

Through the use of bottom-placed transponders and hydrophones on the underwater body, twin screws and transverse thrusters located near the bow, Challenger can maintain position within 150 or 200 feet of a fixed location when subjected to currents of 2 knots and winds of 40 knots.

With respect to drilling capabilities, the longest string which has been suspended was 22,192 feet; the deepest water in which drilling has occurred was 20,483 feet, the deepest penetration that has been made into the bottom was 5709 feet; and the deepest penetration in the basaltic layers below the sediment was 1900 feet. In 1970, Challenger demonstrated for the first time, in 13,500 feet of water, the ability to reenter a drill hole some two feet of diameter–an operation that is now done quite routinely using a sonic beacon for guidance.

Glomar Challenger's contribution to the Deep Sea Drilling Project have been noted widely [6]. Analysis of the cores that have been recovered were key factors in developing models currently accepted by geologists for sea-floor spreading and the plate tectonic theory of the earth's crust.

Glomar Explorer (Figures 2 and 3)

The capabilities of this heavy lift ship, delivered in 1973, are still at the leading edge of the state-of-the-art of deep seabed operations. Designed by Global Marine Development Inc. (GMDI) of Newport Beach, CA (a wholly owned subsidiary of Global Marine Inc.), the ship was constructed by Sun Shipbuilding Corporation of Camden, NJ under the supervision of GMDI, who then operated the ship for the government until inactivation in February 1977. Subsequently, in June 1978 GMDI chartered the ship from the government, and is now operating Explorer as a deep sea mining evaluation platform for a private consortium interested in recovering manganese nodules from the sea floor in the southwest Pacific at depths of about 15,000 feet.

To appreciate the capabilities for working on the deep sea floor, one must look at Glomar Explorer as one element of a total system comprising:

1. ship;
2. subsurface equipment carrier; and
3. construction barge.

Each will be described briefly.

Figure 2. Glomar Explorer.

The Ship [7]

The principal characteristics are:

Length overall	619 ft, 4 in.
Breadth	115 ft, 4 in.
Height to derrick top	263 ft*
Draft at summer freeboard	46 ft, 8 in.
Displacement at summer freeboard (well closed)	63,000 light tons
Cruising speed	10 knots

Explorer' s outstanding feature is the 20,000,000-lb capacity derrick; there is none of equal capacity. Its payload capacity, when one deducts the weight of pipe string, handling gear and subsurface equipment nets out at 8,500,000 lb at a depth of 17,000 ft. Figure 4 shows where this capability fits into the marine spectrum.

As in the case of Challenger, Explorer is fitted with an automatic station keeping system, whose principal features are outlined in Figure 5. It is, however, substantially more advanced not only providing greater accuracy

*Requires removal of derrick top to clear Golden Gate and Bay Bridges in San Francisco Bay.

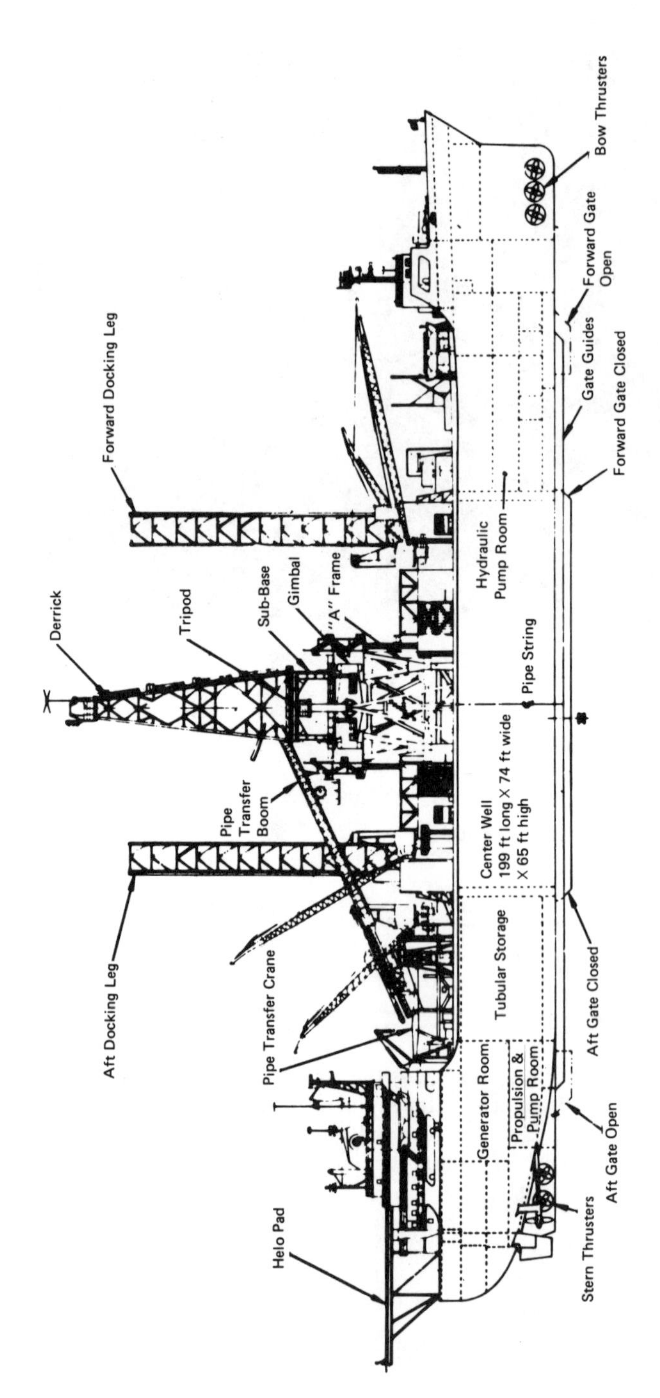

Figure 3. Glomar Explorer–inboard profile.

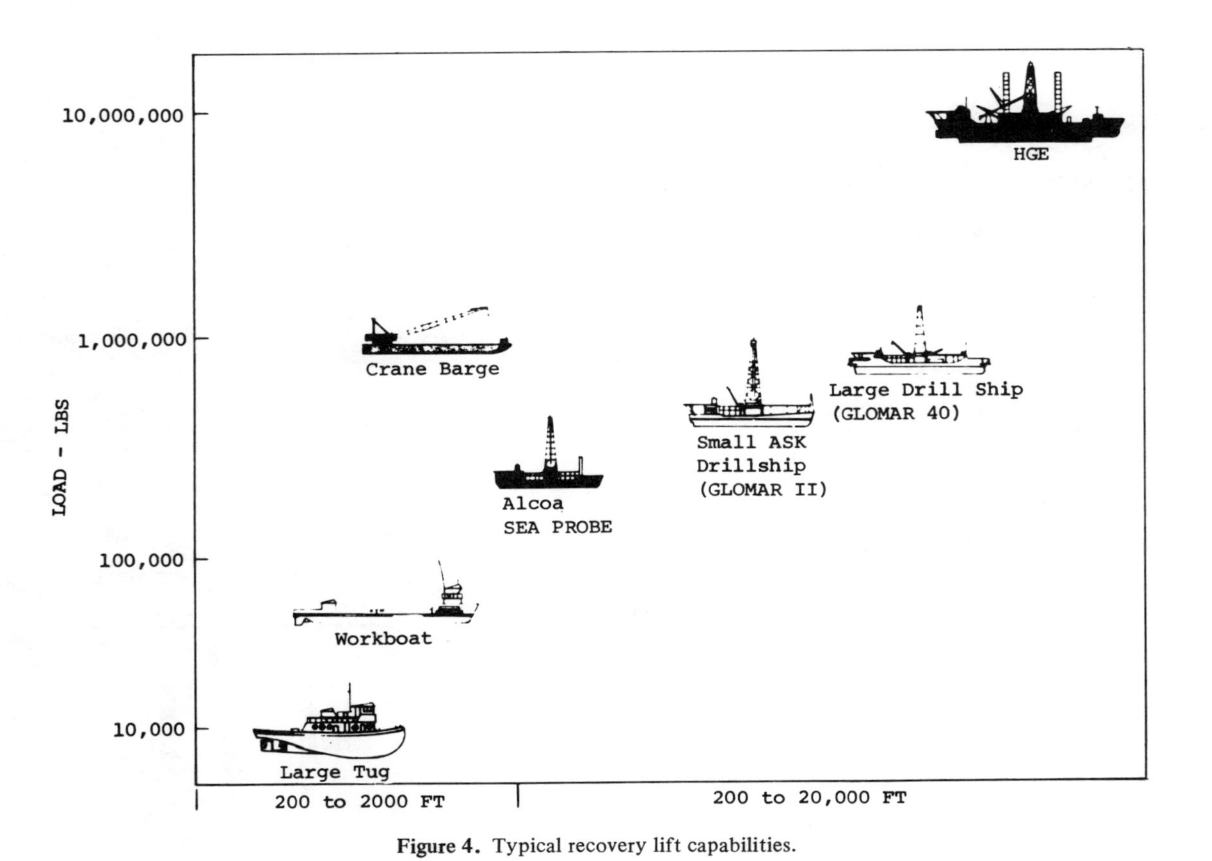

Figure 4. Typical recovery lift capabilities.

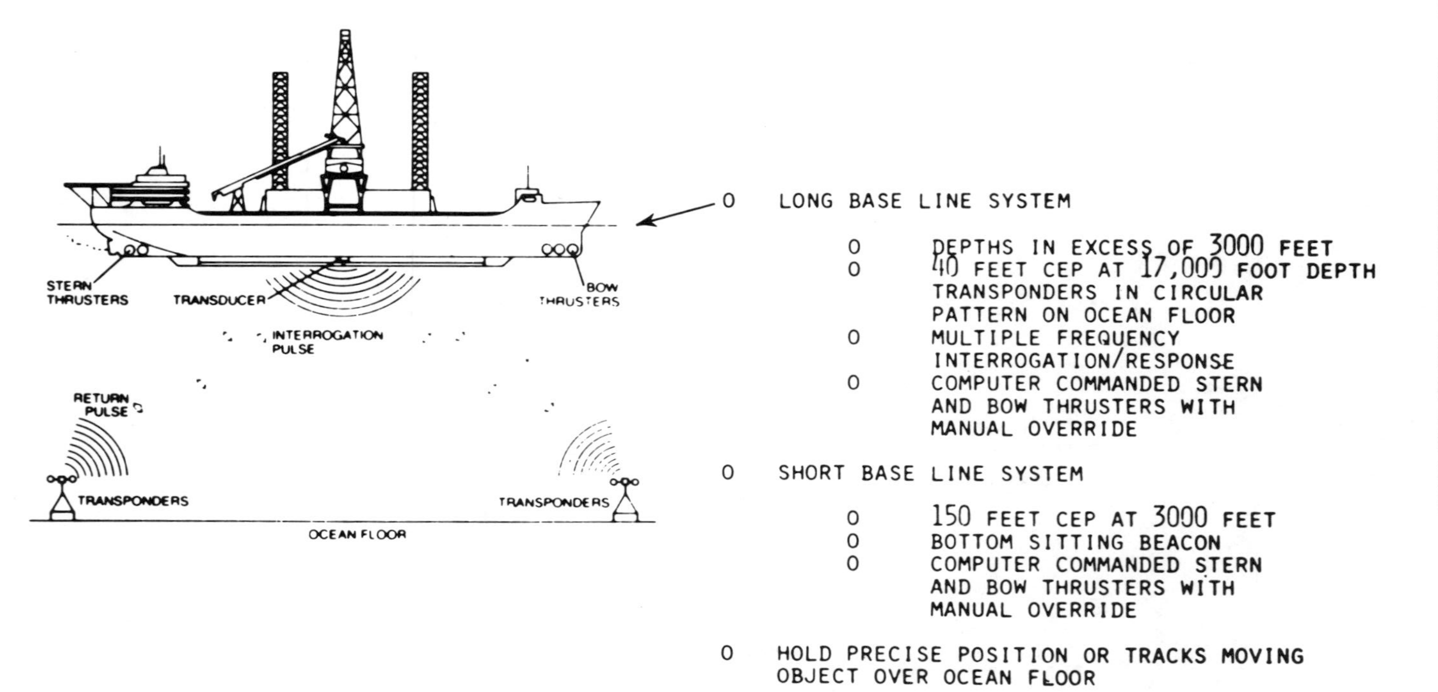

Figure 5. Automatic station keeping system.

in positioning the ship with its long base line system within ±40 feet at 17,000-ft water depth, but will also automatically maintain the heading of the ship within ±2° of that desired with both the long and short baseline systems.

Explorer's unique center well and stable platform are depicted in Figure 6. The gimballed and heave-compensated stable platform provides stable support for the hoisting system. By providing for single amplitude excursions of ±5° in pitch, ±8.5° in roll, and ±7.5 ft in heave, minimal bending moments are induced by ship motions in the highly stressed lifting pipes. The gimbal bearings are especially noteworthy. Having a diameter of 10 ft and fitted with triple races, they are the largest roller bearings ever manufactured.

Explorer's docking legs, located, as shown in Figure 7, at each end of the center well, are used to transfer equipment stowed in the dry well from the well to or from the lift pipe. The docking leg length permits the transfer to be made well below the waterline (100 ft), where relative motions are minimized. The docking leg system can also be used to transfer subsea equipment, such as the subsurface equipment carrier, from the submersible construction barge. The Le Tourneau docking legs used are similar to those used on Le Tourneau offshore oil platforms that are self-elevating ("jack-ups").

The Subsurface Equipment Carrier (Figure 8)

The second element of the Explorer total system is known by this title, but is frequently called simply a "strongback." Its outside dimensions enable it to be stowed in Explorer's center well, which is 199 X 74 X 65 ft. When deployed, it is supplied with about 3000 hp of energy for doing work on the sea floor by two means–hydraulic power furnished from the ship through the hollow pipe string and electric power fed through an electromagnetic cable clamped to the outside of the pipe string. With this energy, the equipment carrier can operate its own thrusters (not shown), its sensors and cameras, and "do work" as required through mechanisms attached to its frame.

Construction Barge (Figure 9)

The element that completes the EXPLORER system is the construction barge, and is designated as the "HMB-1" for "Heavy Mining Barge No. 1." It is essentially a fully submersible floating drydock, and its function is to install and remove subsea equipment in and from the Explorer's center well. Because Explorer's top side is occupied by derrick, docking legs and pipe handling equipment, main deck hatches for access to the well for large items of equipment could not be provided. Hence, the construction barge loads such items (i.e., the subsurface equipment carrier) at a pier, is towed to the Explorer's mooring, is itself submerged (by remote control from an auxiliary work barge) and positioned (Figure 10), directly under Explorer's open well. The subsurface equipment carrier is then lifted by explorer's derrick into the center well with the assistance of Explorer's docking legs. When the

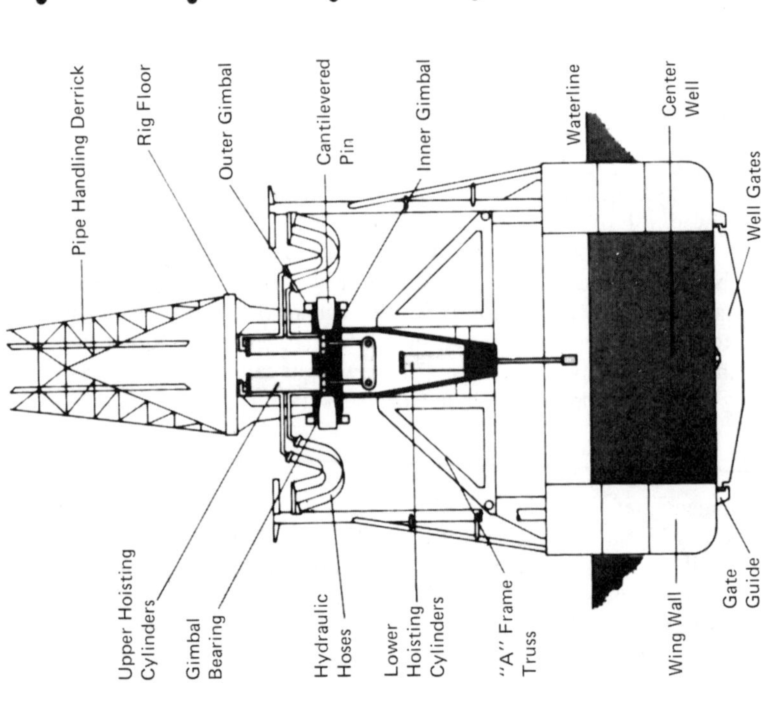

- Gimbaled & Heave Compensated Stable Platform:
  - Isolates Subsea Equipment from Ship Motion
  - 8.5 Million Pound Payload
- Supports –
  - Hoisting Cylinders
  - Rig Floor
  - Derrick
- Stabilized by
  - Inner (Pitch) Gimbal
  - Outer (Roll) Gimbal
  - Pneumatic (Heave) Compensator Rams
- Wet/Dry Center Well
  - Secure Dock Area
  - 199′ × 74′ × 65′
  - Horizontal Sliding Gates

Figure 6. Stable platform and center well.

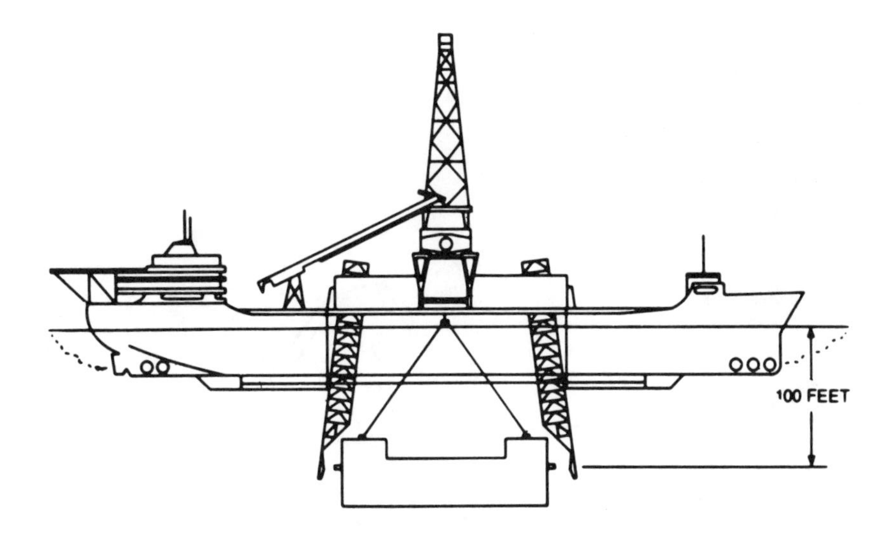

Figure 7. Docking leg system.

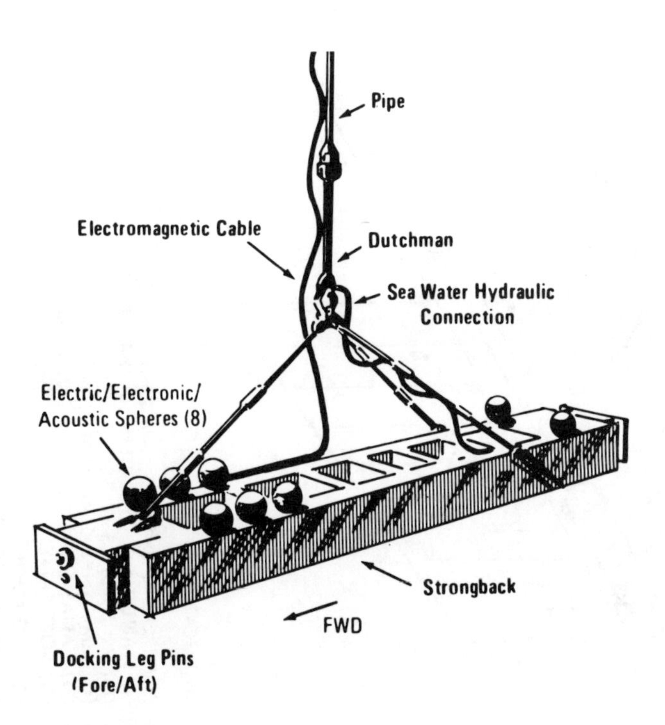

Figure 8. Subsurface equipment carrier.

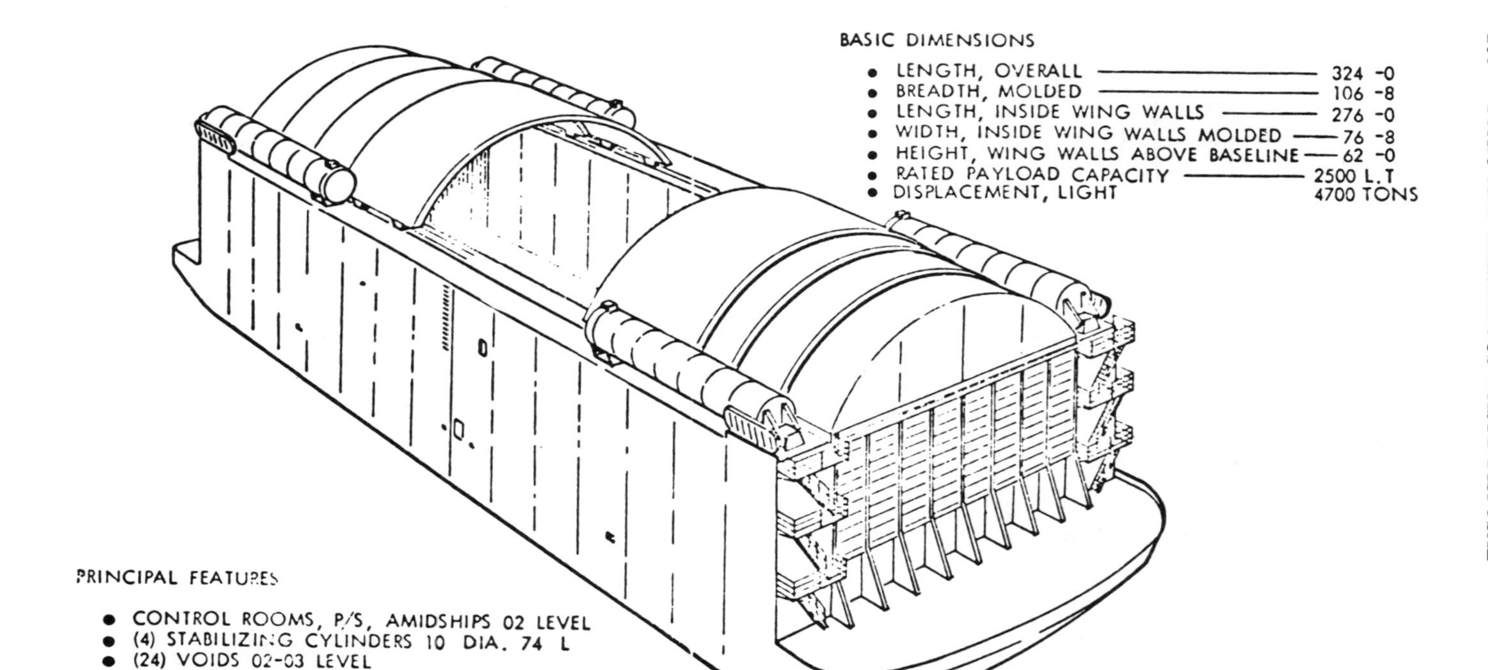

Figure 9. Construction barge (HMB-1).

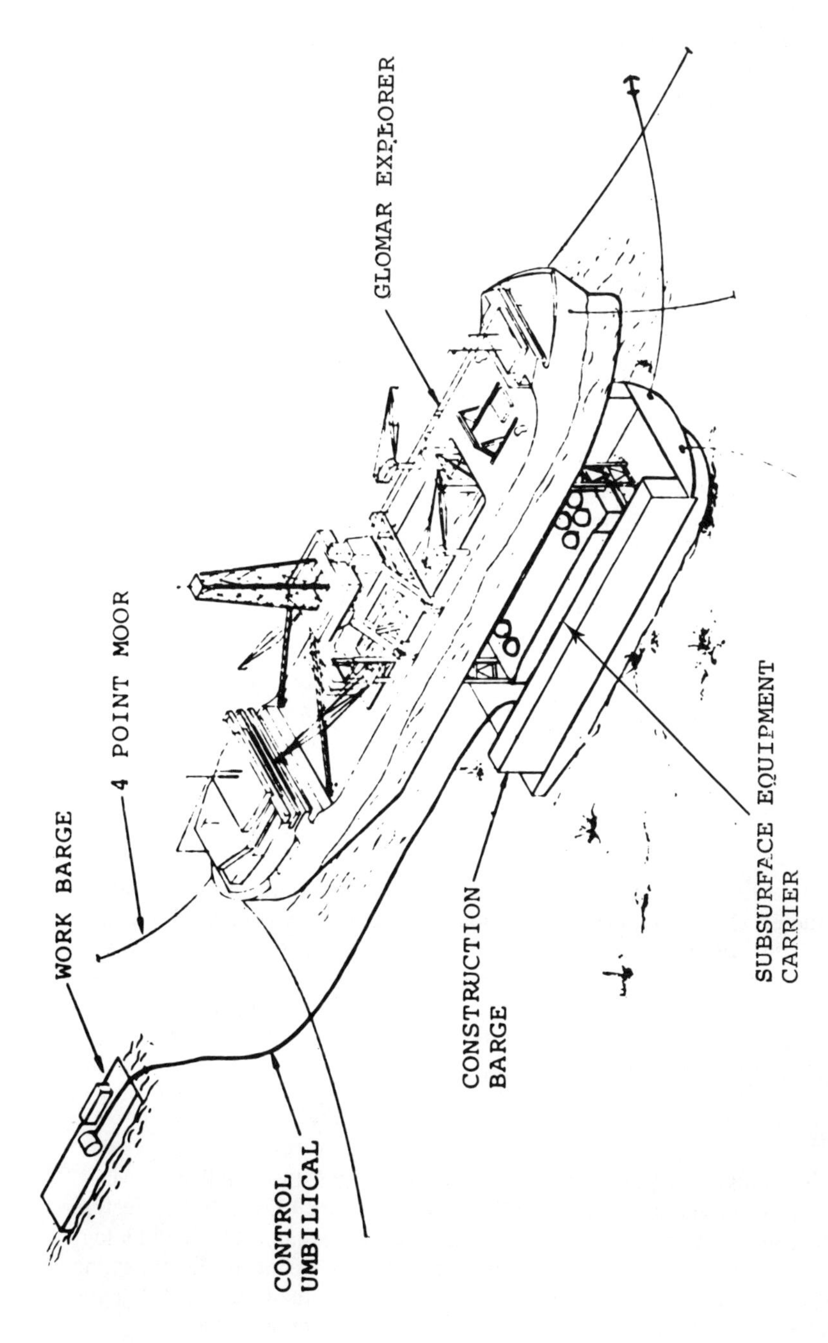

Figure 10. The Glomar Explorer system.

transfer is completed, the well gates can be closed, the construction barge refloated and towed to base, and Explorer can depart from the mooring with the subsurface equipment carrier securely stowed in its well and ready for subsequent operations.

SEABED DISPOSAL OF HLW/SPENT FUEL AND TRU

The Multibarrier Concept

Seabed disposal of HLW/spent fuel as well as TRU waste is based on a multibarrier concept that would contain radioactive waste long enough for it to decay to harmless background levels. As shown in Figure 11, it is essentially a system of barriers comprising the waste form, the waste container, any controlled modification of the emplacement media, the benthic layer and the water column. As noted earlier, this concept is focused on the consolidated, fine-grained, red clay, deep-sea sediments whose sorption and diffusion properities promise a 10^6-10^{13}-year barrier to radionuclide migration toward the hydrosphere. The half-lives of the major constituents of radioactive waste are:

Radionuclide	Half-Life (Yr)
Americium-241	460
Cesium-135	2×10^6
Cesium-137	30
Iodine-129	1.6×10^7
Plutonium-239	2.4×10^4
Radon-226	1600
Strontium-90	28
Thorium-230	7.6×10^4
Tritium	13

Figure 11 shows that the sediment being considered as the principal barrier could provide the requisite protection. If such wastes were placed directly on the sea floor (in the water column), manmade barriers or waste form/containers could not be expected to survive long enough to prevent entry of radionuclides directly into the biosphere. Thus, researchers in this field have concluded that the water column itself cannot be considered a primary barrier in the containment system [5].

Sites

The most favorable sites are Atlantic and Pacific abyssal hill and swale regions of midplate, midgyre (MPG) areas. These are 15,000-20,000 feet deep, and are climatically and geologically stable, economically and biologically sterile, and secure from inadvertent retrieval. Figure 12 illustrates the plates which form the earth's crust, and the stars in the Pacific and Atlantic plates show candidate MPG sites having 60-100 ft sediment thickness. The

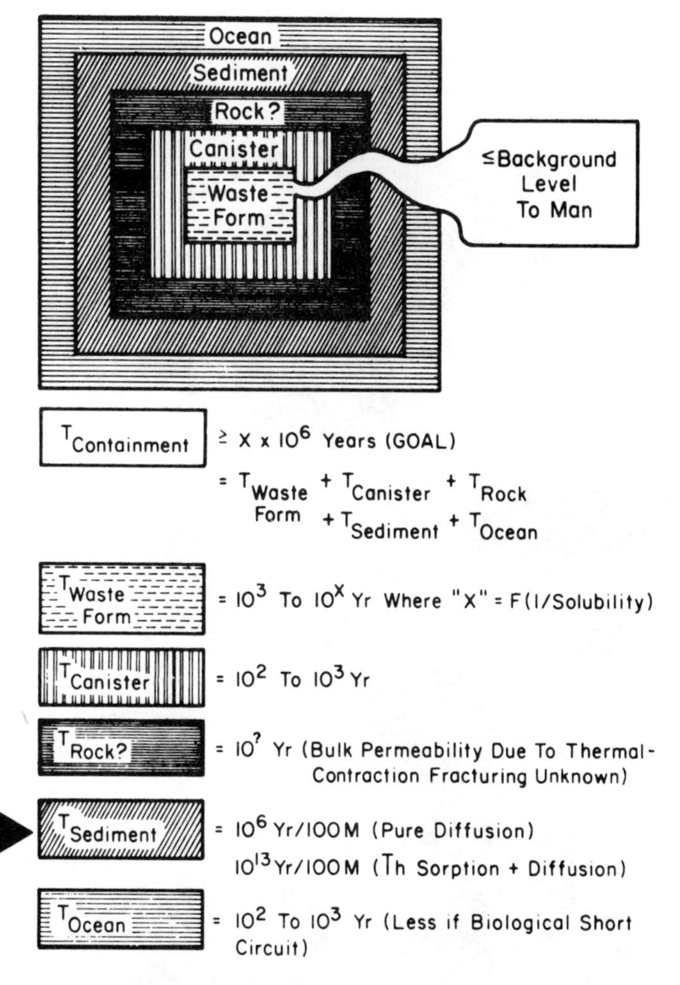

Figure 11. Multibarrier concept [5,8].

abyssal hills in these regions have about 75% as much sediment cover as the valleys, indicating some downslope concentrations. The reader will recall that Glomar Challenger's core retrieval work under the Deep Sea Drilling Project was the primary scientific tool that developed the data on which the plate tectonic theory was formulated.

Development Needs

Seabed disposal shares with other concepts the handicap that its effectiveness cannot be demonstrated in advance—that is, the ability to isolate high

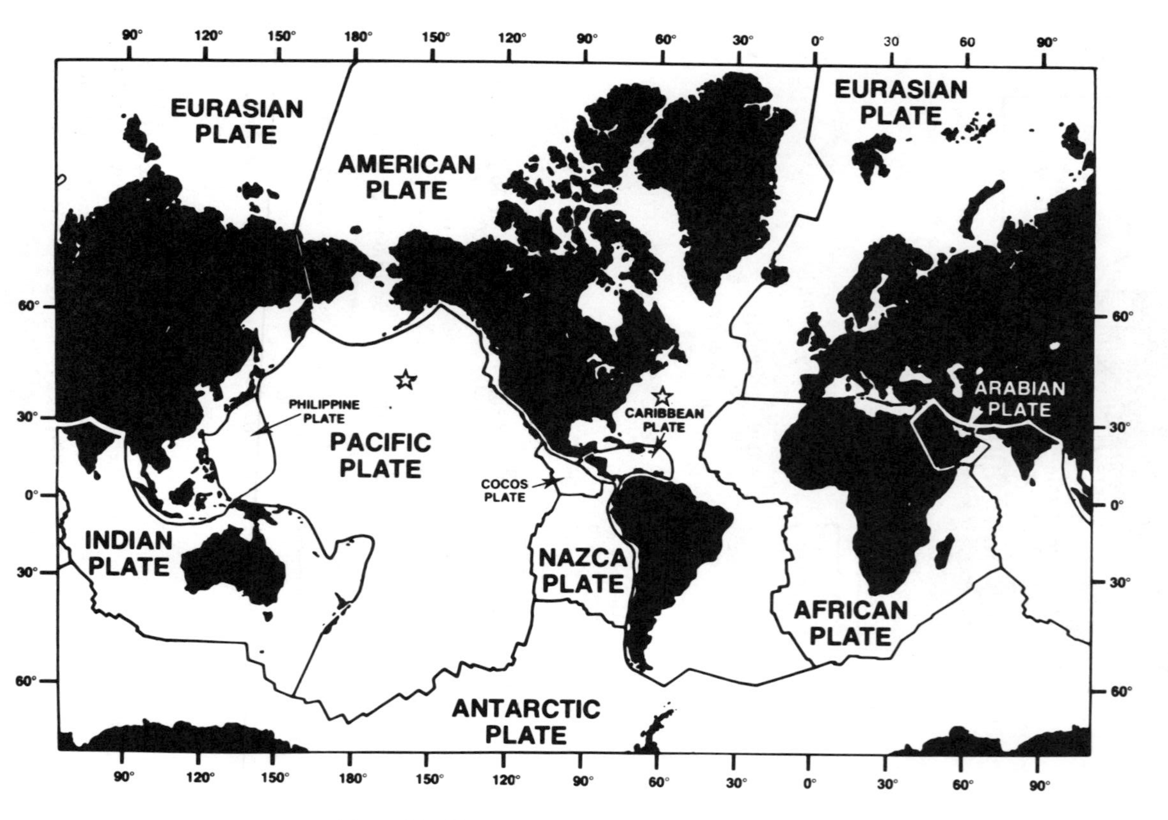

Figure 12. Plate tectonic map of the world [8].

level wastes from the biosphere for some 10^6 years is simply not demonstrable in our generations. Thus, an analytical model must be developed and its acceptability established by demonstration of the validity of its key elements by laboratory-and full-scale tests. In such an undertaking it is important that scientific investigations such as identifying the best sediment for retention of nuclides and the disposition of sediments as to content and thickness be carried out in parallel with ocean engineering aspects. In the latter case, emplacement, monitoring and retrieval techniques need to be developed concurrently. Unfortunately, the DOE program is so fund-limited (less than 5% of that for mined repositories) that seabed disposal assessment is now being carried out through a series approach: (1) determine environmental feasibility; (2) assess engineering feasibility; and (3) provide demonstrations of concept. This necessarily involves a long time for accomplishment, probably 20 years at the current rate.

Thermal Release

The response of the sediments to heat generated by a waste container has been difficult to determine in a laboratory in which, for example, the hydrostatic pressure on the MPG seafloor (about 500 atm) cannot conveniently be duplicated in an appropriate test. An appreciation for the heat release problem, which applies to all concepts, can be gained from noting that a proposed standard steel canister (10 ft long X 12 in. diameter) will release 10-30 kW of heat when newly filled. An in situ verification test in seabed sediments is clearly needed to achieve an understanding of the phenomena. Figure 13 illustrates an experimental concept developed by GMDI for carrying out such a field test—certainly not a small undertaking at 17,000 ft, yet essential to concept validation.

Emplacement

A freefall penetrometer technique has been considered for canister emplacement. Laboratory tests [8] have shown that high-speed projectiles fired from a compressed air gun into varying thicknesses of sediment were followed by immediate and total hole closure. Slow sediment penetrations created cavities whose walls flowed gradually inward. A severe shortcoming of the high-speed penetrometer approach is its uncertainty as to final location, which would make monitoring difficult and retrieval probably impracticable. Use of dynamically positioned drilling ships, as previously described, would permit fully controlled emplacement, monitoring and retrieval. Figure 14 illustrates the type of emplacement site that is visualized. Validation of this technique would entail in situ tests of the effectiveness of the hole closure by sediments emplaced by wall cavity collapse, by jetting or other means. Figure 15 illustrates another emplacement approach that contemplates use of the unique capabilities of a ship such as the Glomar Explorer for placing on, or jetting into the seabed, a large concrete crypt, or "tube" rack, for radioactive waste canisters.

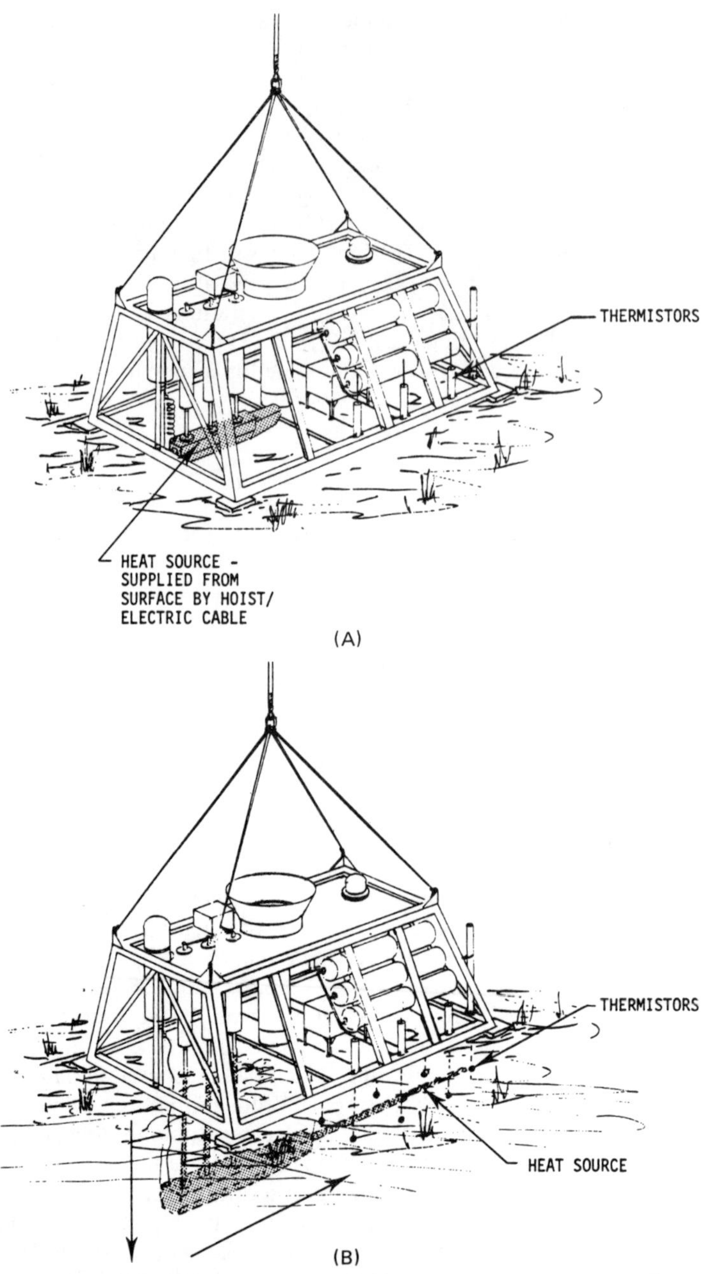

Figure 13. HLW–in situ heat transfer test in seabed. (A) Test jig on sea floor. (B) Thermistors deployed: heat source thrust vertically into sediment and horizontally to center of thermistor array.

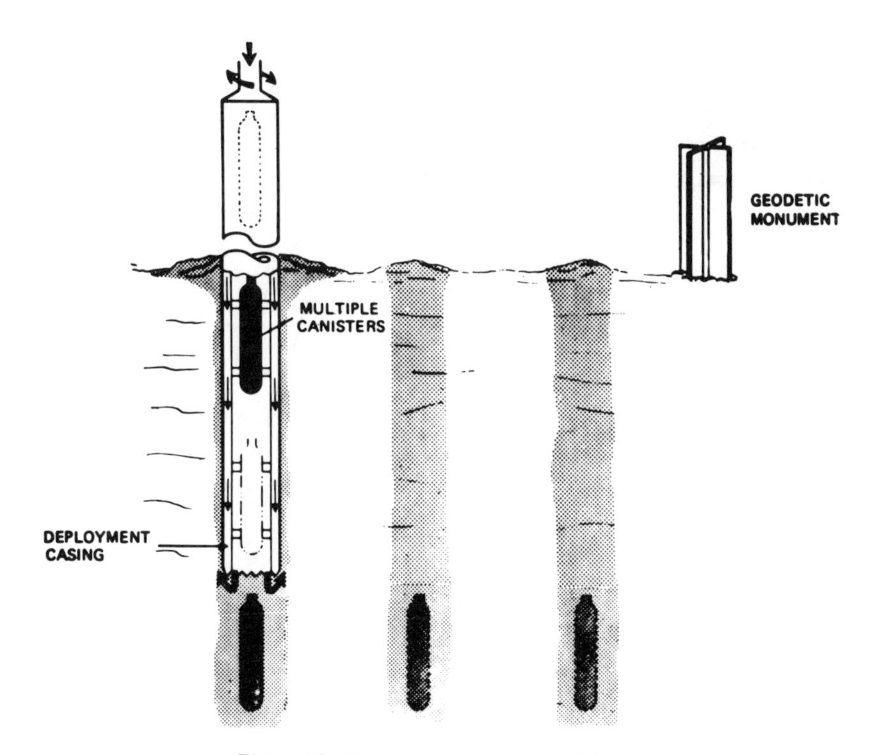

Figure 14. Canister placement within sediment.

SEABED DISPOSAL OF LLW AND MILL TAILINGS

Seabed disposal for these wastes represents a much simpler problem than HLW/spent fuel because of the absence of need for shielding and the absence of thermal release. Although the U.S. discontinued ocean disposal in 1970, the U.K., Belgium, the Netherlands and Switzerland have been dumping solid wastes under the control of the Nuclear Energy Agency (NEA) since 1967 in the deep North Atlantic. The International Atomic Energy Agency (IAEA), which was entrusted with recommending methods and criteria for radioactive wastes by the London Dumping Convention of 1972 [9], is now developing new definitions and criteria based on radioactivity release rates [10]. The U.S., together with IAEA, is moving to agreement on the proposed revisions. The U.S., however, is dedicated to disposal of LLW wastes on, in or under the seabed to isolate the wastes from the biosphere in a manner that will be environmentally acceptable during the period of expected radioactivity ($10\text{-}10^2$ yr–not 10^6 yr).

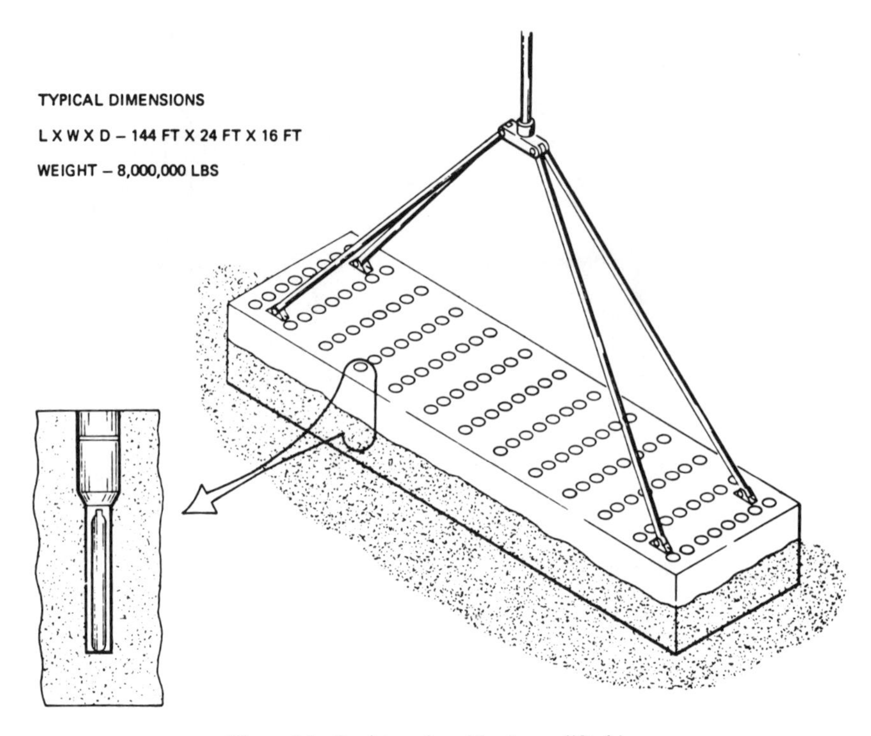

Figure 15. Canister placed by heavy lift ship.

EPA Program for LLW

The U.S. Environmental Protection Agency (EPA), under its statutory responsibility [11] for issuing permits for ocean disposal of all wastes, is embarking on a development program that is expected to lead to ocean disposal of LLW by the mid-1980s. EPA surveys conducted since 1974 of former ocean dumping sites off the East and West Coasts indicate excellent potential for seabed emplacement of LLW in canisters that would overcome the plaguing difficulties being encountered in land burial. As a result of these surveys, EPA concluded:

(1) Techniques formerly used to package the radioactive wastes for ocean disposal were, in general not adequate to insure that the wastes would remain isolated from the surrounding environment.

(2) If ocean disposal of LLW were to begin in the future, the technology currently exists to precisely monitor a deep ocean site to detect the possible release and movement of selected radionuclides and to recover waste packages disposed at depths up to 2,800 m [11].

Mill Tailings

Large quantities of mill tailings, currently 140,000 tons and expected to increase to 370,000-485,000 tons by the year 2000, are now recognized as hazardous because of their slow release of radon. Shielding and thermal release are not issues in this case, but isolation from the biosphere is needed. Figure 16 illustrates a concept for slurrying such wastes, after grinding to convenient grades, from a barge to large holes in the sea floor (2000 ft depth X 4 ft diameter) that could be created by a conventional drilling ship,

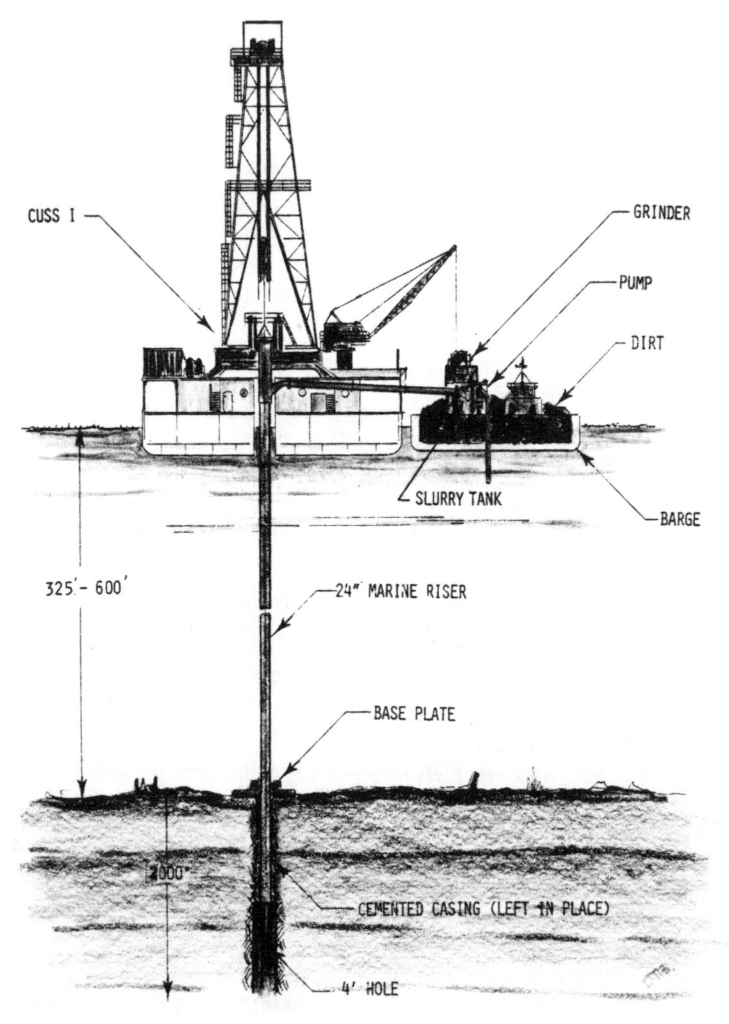

Figure 16. A concept for disposal of mill tailings.

such as the Cuss I that is shown in the sketch. Delivery of slurried mill tailings from the drilling vessel to the sea floor would be through a conventional marine riser that would be left in place after the hole drilling is completed so that there would be no ocean contamination during slurry transfer. When the hole is almost filled with slurried mill tailings, a cement cap would be produced by the drilling ship using its conventional cementing capabilities. Thus, a relatively simple technique is now available for isolating large quantities of mill tailings or other types of LLW of similar characteristics by the use of available deep-sea technology.

LLW vs HLW

The major differences between disposal of LLW and HLW/spent fuel lie in the increased efforts involved in achieving proper emplacement and the thermal dissipation phenomena for the latter. It appears, therefore, that seabed disposal of LLW should logically precede HLW. Thermal effects of HLW could be examined concurrently with LLW emplacement and, if successful, could provide an excellent basis for developing HLW technology and gaining domestic and international acceptance.

SUMMARY AND CONCLUSIONS

The data presented in this chapter may be summarized as follows:

1. The quantities of LLW, TRU, HLW/spent fuel and mill tailings are significant and are increasing.
2. The seabed alternative is the most promising of nonconventional techniques.
3. Deep-sea technology is advancing rapidly and already provides excellent tools and techniques for seabed disposal.
4. Deep-sea sediments appear to provide the requisite barrier to radionuclide migration, but additional scientific research on its disposition and its physical and dynamic responses to emplacement is required.
5. Ocean engineering for in situ tests and development of emplacement techniques are also needed in parallel with the scientific investigations in order to achieve strong synergism among them.
6. Seabed disposal of LLW and mill tailings is simpler than HLW/spent fuel, and development of the former would enhance the latter.

The potential advantages of seabed disposal include remoteness from present human activities, the high isolation capabilities of the sediments, the high heat sink capability of the oceans, the large areas available and the possibility of avoiding the problems of finding and gaining access to adequate land sites. Some of its potential complications are the aspects of monitoring, the requirement for transport to ports and ocean transport, special port and ship facilities, and the need for national and international acceptance.

On balance, the prospects for seabed disposal of radioactive wastes are positive, and it should be considered a prime alternative to mined repositories.

REFERENCES

1. "Issues Pertaining to the Scientific and Technical Knowledge Relevant to Geologic Disposal and Isolation of Nuclear Wastes," Executive Office of the President, Office of Science and Technology Policy (8 June 1978).
2. "A Classification System for Radioactive Waste Disposal–What Wastes Goes Where?" NUREG 0456, FBDU 224-10 (June 1978).
3. "Sub Group Report on Alternative Technology Strategies for Isolation of Nuclear Waste–TID28818," Interagency Review Group on Nuclear Waste Management, and Report to the President by the Interagency Review Group on Nuclear Waste Management; TID-28817 (October 1978).
4. "Report of Hearings Before Sub-Committee of the Committee in Government Operations–Low Level Radioactive Waste Disposal," HOR, Feb. 23, March 12 and April 6, 1976.
5. "Seabed Disposal Program, Annual Report–Part I, Jan-Dec 1976," Sandia Laboratories (October 1977).
6. Nierenberg, W. A. "The Deep Sea Drilling Project After Ten Years," *Am. Scientist* (Jan-Feb 1978).
7. "The GLOMAR EXPLORER–Deep Ocean Working Vessel, Technical Description and Specification," Global Marine Development Inc., Newport Beach, CA (1975).
8. "High Level Nuclear Wastes in the Seabed," *Oceanus* 2(1) (1977).
9. The Convention on the Prevention of Marine Pollution by Dumping of Wastes and Other Matter (1972).
10. Memorandum by Director General, Board of Governors, International Atomic Energy Agency, File No. GOV/1889 (27 April 1978).
11. "Ocean Dumping in the United States–1977," Fifth Annual Report of the EPA on Administration of Title I, Marine Protection, Research and Sanctuaries Act of 1972 as amended (March 1977).

CHAPTER 17

DEEP-WELL DISPOSAL: A VALUABLE NATURAL RESOURCE

Ray W. Amstutz
Williams Brothers Engineering Company
Tulsa, Oklahoma

Deep-well disposal is the emplacement and storage of unusable liquid wastes in the pores of deep subsurface geological formations which already contain unusable saltwater.

The term "deep-well" as used in this text is defined as a well which has been drilled to a geological formation far below and isolated from all freshwater strata. More than half of the disposal wells now in operation are 2000-6000 ft (610-1800 m) deep. Some are deeper, ranging to 12,000 ft (3700 m) or more. A few are shallower than 2000 feet but they generally are permitted only where unusual geological conditions exist.

Properly planned and implemented, deep-well disposal is an environmentally acceptable method of storing liquid wastes which are economically unusable in today's complex industrial operations. The basic principle which cannot be compromised is that the injected waste must not present any danger to potable or other usable waters, other natural resources or human surroundings. Usable waters as defined by the U.S. Environmental Protection Agency (EPA) and state regulatory agencies include waters with a total dissolved solids (TDS) content of less than 10,000 mg/1, i.e., waters which are up to 30% as salty as seawater.

WHEN OR WHERE TO USE IT?

Deep-well disposal should be considered only when the liquid wastes cannot be treated or disposed of economically in other ways. The fundamentals of deep-well disposal are those of geology and the nature of the waste. The industry which is generating the waste must be located over a suitable geological formation.

Figure 1 is a map of the 48 states showing the geographical areas which are classified as favorable, possible and unfavorable. Locations of many of the existing disposal wells also are shown. In many cases the circles depicting well locations represent more then one well. There are areas in 36 of the 48 continental states which are underlain by sedimentary strata which are geologically attractive for deep-well injection. Six additional states have limited geological conditions with possible locally favorable conditions. The remaining six states have few or no areas that are suitable.

The areas classified as favorable have a wide range of attractions. Certain areas have geological and economic advantages. As the industrial complex continues to expand, we may be forced to use the superior deep-well areas to dispose of wastes that are generated in less favorable areas.

In the future it may be common to have wastes moved long distances from plant sites just as raw materials now frequently are moved long distances to plant sites. For example, a pipeline now carrying natural gas to an industry may be paralleled by another pipeline carrying liquid waste in the opposite direction. In many cases, liquid wastes now are being moved by pipeline, trucks, railroad and barge to offsite disposal operations. Deep disposal wells will be a significant factor in the ultimate disposal of liquid wastes. Their capacities are enormous, making them a valuable national resource.

INDUSTRIAL ACCEPTABILITY

Disposal of oil field brines (dissolved solids concentrations of 100,000-200,000 mg/1 are common) into injection wells was becoming a common practice in the 1930s. There was—and still is—no other feasible method of disposing of these enormous quantities of heavy brines in the nation's inland oil fields.

Initial regulatory control was minimal. Many small operators had little or no engineering capability so poor practices resulted in environmental damages. However, the environmental damage was small compared with the common abuses resulting from poor surface disposal practices in other industries at that time.

Improved disposal well completions, operational procedures and regulatory controls soon raised the level of dependability to the point that environmental damage became a rare occurrence. Approximately 40,000 injection wells now are in operation in the nation's oil fields. Total brine injection is more than 10 million bbl/day (300,000 gal/min or 19 m^3/sec).

Industries other than the oil-producing industries began using deep disposal wells for hazardous liquid wastes in the 1950s. There are about 300 industrial disposal wells in the nation today.

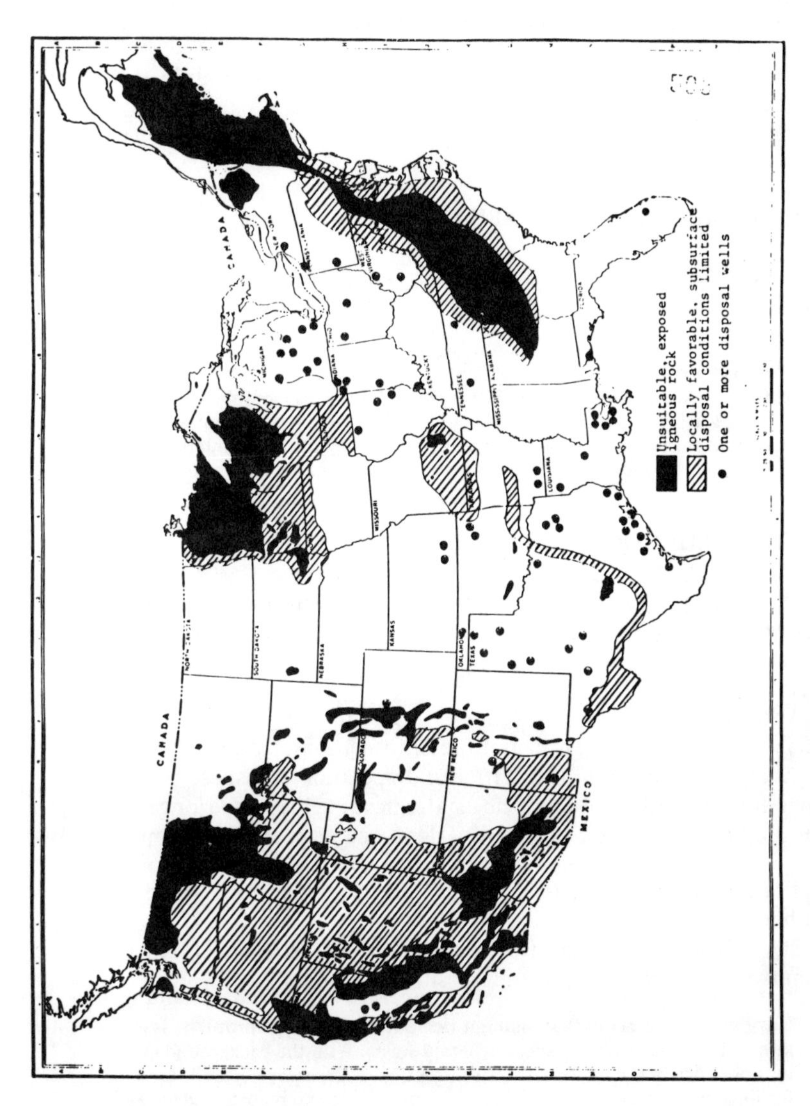

Figure 1. Geographical areas classified for suitability for deep-well injection.

Some industries now successfully using deep well disposal are:

airlines	glass	paper
auto manufacturing	metal plating	petrochemical
chemical	minerals	pharmaceutical
drug	miscellaneous manufacturing	space
fertilizer	munitions	steel
food	oil refining	textile

Total disposal rate in industrial disposal wells is in excess of 28,000 gal/min (1.8 m^3/sec).

ENVIRONMENTAL ACCEPTABILITY

Use of deep disposal wells now has less opposition from environmental groups for one or more of the following reasons:

1. Alternative methods of hazardous waste disposal often are not feasible or entail a greater risk of environmental damage.
2. Disposal wells permanently remove the liquid wastes from the environment.
3. They are becoming more knowledgeable about: (1) the concept of deep, underground storage; and (2) the extensiveness of controls and regulations that are imposed on industry through both state and federal regulatory programs.

STATE REGULATIONS

State regulatory programs vary widely. States which have no suitable sedimentary geological formations obviously prohibit deep-well disposal operations and have no need for regulatory controls on installations and operations.

At the other extreme, states such as Texas, Illinois, Louisiana and Oklahoma, which have excellent disposal formations, have extensive requirements and regulations for installation and operation of disposal wells. These states and most other oil-producing states have engineers and/or geologists who are familiar with the concepts of deep-well disposal. When appropriate compliance with regulations is assured, they approve deep-well disposal applications and in some cases, strongly encourage the use of this disposal method.

The following quotations from publications of state regulatory agencies from two different states illustrate positive positions:

> It is the belief of many that underground injection, if done properly, is a safe and sensible way to dispose of liquid waste. With the background gained by this office in the use of salt water injection wells, as well as wells used for injection in secondary recovery projects, all with no reported adverse effects, has prompted the consideration of approval of waste disposal wells.

This state has 57 active wells, with an average disposal rate of 110 gal/min (7 liter/sec) per well.

> The advantages of underground disposal of waste are: (1) the fate of the waste is, in general, known and understood; (2) the waste is contained and can be isolated from man's food, water, air, and activity; and (3) the waste can be recovered if the need arises.

This state has 91 active wells with an average disposal rate of about 100 gal/min (6 liter/sec) per well.

FEDERAL REGULATIONS

Prior to passage of the Safe Drinking Water Act, all regulations and controls for deep-well operations were enforced by the states. The Safe Drinking Water Act provided the EPA with authority to require minimum standards and safeguards for industrial disposal wells.

The more recent Resource Conservation and Recovery Act (RCRA) provides EPA with a comprehensive underground injection control program. This act authorizes EPA to require state regulatory agencies to provide the necessary controls to assure protection of all underground water reserviors containing water with less than 10,000 mg/l TDS. Exceptions may be granted if there is no possibility of the water being used or needed as drinking water in the foreseeable future.

The requirements imposed upon the states are extensive. A few of the many requirements are listed below.

1. maintain an inventory of all disposal wells;
2. use court action against violators;
3. inspect, sample and monitor operations;
4. require operators to keep appropriate records and submit periodic reports;
5. use appropriate procedures for issuing installation and operational permits;
6. review annually the state's underground injection control program to assure compliance with federal regulations;
7. review all well permits and make tests to prove mechanical integrity of each well at least once every five years; and
8. require appropriate responses to a long list of specific questions regarding hydrology and geology of the area, characterization of the waste streams, expected rates, pressures, well completion methods, preinjection treatment requirements, calculated zone of influence, etc.

ACCEPTABLE WASTE LIQUIDS

A wide range of liquid wastes are suitable for deep well disposal. Some examples are:

- dilute or concentrated waste acids
- weak or strong alkaline solutions
- solutions of heavy metals
- inorganic solutions
- hydrocarbons, including chlorinated hydrocarbons
- toxic and other hazardous solutions
- organic solutions including those with high biochemical or chemical oxygen demand (BOD or COD)
- solvents

The tolerance of deep-wells for waste solutions is very wide, but the tolerance for suspended solids (SS), in most cases, is very narrow. Removal of SS before injection is a requirement for successful long-term disposal in most subsurface strata. A second requirement is that the viscosity of the waste solution be reasonably close to that of water.

Because suspended solids usually are detrimental to injectivity, it follows that the waste liquid must have sufficient chemical stability to permit injection before significant precipitations occur. In certain cases, compatibility of the waste liquid with the host disposal formation also can be a problem.

INSTALLATION REQUIREMENTS

An investigation of alternative disposal methods is required before deep-well disposal will be considered for approval. If no alternative methods are feasible, studies then must be conducted to determine regulatory, geological, chemical and technical feasibility of deep-well disposal.

To obtain a permit for the drilling of a test well, an engineering design and drilling prognosis must be prepared to assure completion of the well in a manner that will meet all regulatory specifications.

Figure 2 is a schematic of a typical disposal well. The surface casing must extend to a depth below all freshwater zones and be cemented from the bottom to the surface. The next string of casing must extend to the selected disposal formation and it too, must be cemented from the bottom to the surface. A third string of pipe, the injection tubing, must extend to the bottom of the casing and the annulus sealed with an appropriate packer.

These requirements provide a very conservative barrier between the disposal liquids and the freshwater zones. It consists of three concentric strings of pipe, two cement sheaths and a sealed (and monitored) annulus. These requirements may represent some "overkill" but they do assure the integrity of our subsurface freshwater zones and probably provide considerable comfort to the regulatory personnel and the operator of the well.

During the drilling of the well, extensive hydrological, geological and lithological data must be gathered, including:

- thickness and depth of the disposal formation and of the confining aquicludes above and below the disposal formation
- lithology of the disposal formation
- permeability and porosity of the disposal zone and the confining beds
- water saturations and water analyses
- formation temperature and pressure
- compatibility of the waste liquids with the disposal and confining formations
- compatibility of the waste liquid with the native water from the disposal formation
- expected disposal rate and pressure

If the test data indicate that the proposed disposal well will operate satisfactorily and safely, state approval then must be requested for conversion of the test well to an operational well.

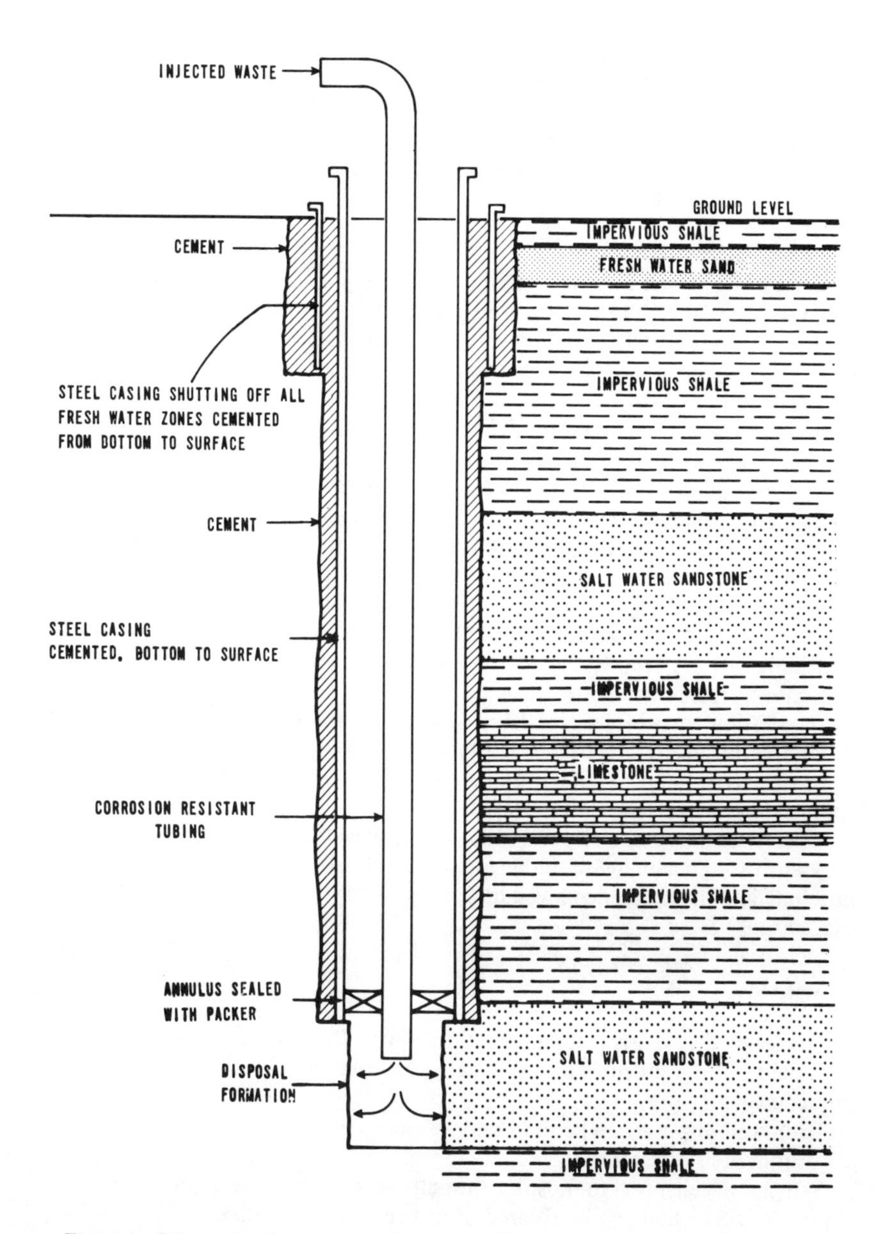

Figure 2. Schematic of typical completion method for a deep disposal well. Typical depth: 1000-2000 m.

Some wastes in their raw form are not suitable for injection. Filtration is the most common treatment required.

Approximately 10% of the industrial disposal wells are taking the desired flowrates under vacuum. The remaining 90% require injection pumps to provide the positive force necessary to displace native fluids and allow movement of the waste into the host formation.

COSTS

Costs of deep-well disposal systems vary widely. Geological conditions and waste characteristics are the principal factors affecting costs. The cost per foot of drilling and completing a well increases with depth. The cost per foot for a very deep well can be up to ten times as much as the cost per foot for a relatively shallow well.

A few waste streams need no treatment or filtration. Others require expensive treatment and filtration. Many wells require injection pressures at the wellhead of several hundred psi, thus necessitating the installation of high-pressure pumps. Large savings in installation and operating costs are effected in those cases where the well accepts the waste under vacuum.

Corrosiveness also is a big factor in cost variability. Use of exotic corrosion-resistant tubing and packer in the well, surface piping, pumps, filters, etc., can increase the cost of materials by an order of magnitude. Availability of drilling rigs and monitoring and testing requirements of the particular state also influence the costs.

Installation costs of deep-well systems usually are less than half of the cost for alternative disposal methods. Costs for deep-well systems typically range from \$200,000 to \$400,000, although they can be somewhat lower or appreciably higher.

Operational costs for disposal well systems usually are low. Automation is extensive, so personnel requirements are minimal. Total costs for installing and operating deep-well disposal systems may be 0.02-0.4 ¢/gal (\$5-100/m^3).

SUMMARY

Deep-well disposal is recommended for those difficult cases where surface treatment is determined to be technically, environmentally or economically impractical.

Deep-well disposal can be used only where existing subsurface geological conditions are satisfactory.

Satisfactory disposal in most but not all cases is limited to liquids that are free of SS or that can be filtered to provide a suitable filtrate.

Extensive studies, testing and planning are required to assure that the deep-well injection system is properly designed to provide the necessary protection to the environment and to obtain the necessary regulatory agency approvals.

When properly designed, installed and used, deep-well disposal systems provide permanent storage for many liquid industrial wastes. (Could some of today's wastes be tomorrow's raw material sources?)

These vast subsurface storage formations contain heavy brines that have been isolated from the earth's ecological sphere for 100-500 million years. The volume of additional liquids added through the deep well disposal process is infinitesimal.

This valuable national resource is providing a satisfactory solution to the problem of waste disposal for numerous industries and is helping them cope with rising environmental costs.

CHAPTER 18

SOIL INCORPORATION (LANDFARMING) OF INDUSTRIAL WASTES

D. E. Ross and H. T. Phung
SCS Engineers, Inc.
Long Beach, California

INTRODUCTION

Industrial wastes are typically disposed of in sanitary and chemical landfills, in deep wells, or by incineration [1]. Until recently, some wastes were simply dumped into waterways.

Land disposal of industrial sludge and wastewater by mixing into the surface soil is called soil incorporation. This method is also known as land farming, land cultivation or sludge farming. Spreading of organic wastes (e.g., animal manure, crop residues and sewage) on agricultural land to supply plant nutrients and organic matter is a practice dating far back in history. However, land farming of industrial wastes is of recent origin, particularly as an alternative to landfilling. In general, the practice involves four basic steps:

1. application of waste onto or beneath the surface soil;
2. mixing the waste with the surface soil to aerate the mass and expose waste to soil microorganisms;
3. adding nutrients or other soil amendments (optional); and
4. remixing the soil/waste mass periodically to maintain aerobic conditions.

Recently, SCS Engineers, Inc. has completed a study on the state-of-the-art evaluation of soil incorporation funded by the U.S. Environmental Protection Agency (EPA) [2]. Since that study, SCS Engineers has investigated the feasibility of implementing soil incorporation operations for several industrial clients. This chapter reports major findings of the EPA-sponsored study and recent industrial work.

WASTE DEGRADATION AND VOLUME REDUCTION

The objective of soil incorporation practices is to ensure that the chemical constituents in the waste are retained in the surface layer and/or decomposed. The important processes that contribute to waste volume reduction include microbial degradation, nonbiological (chemical and photochemical) degradation, evaporation and volatilization.

Microbial Degradation

The principal groups of microorganisms present in surface soils are bacteria, actinomycetes, fungi, algae and protozoa [3]. In addition to these groups, other micro- and macrofauna are often present, such as nematodes and insects. The bacteria are the most numerous and biochemically active group of organisms, especially at low oxygen levels.

The microbial population in most soils constitutes a biochemically complex system capable of producing unique enzymes that can degrade many organic substances [3,4]. The inorganic fraction of an industrial waste may be compatible with soil microorganisms but would not be attacked. These microorganisms normally adhere to the surfaces of the soil colloids. Most groups are heterotrophic, deriving energy from the breakdown of organic substances.

In general, conditions favorable for plant growth are also favorable for the activity of soil microorganisms. Environmental factors influencing plant growth and the range of values conducive to growth are [2]:

- soil pH–range of 4-10; pH 7 is generally optimal
- air temperature–varies with type of organism; 86-104°F (30-40°C) is optimal.
- soil moisture content–a certain minimum water content is required, generally in the range of 30-90% of the water holding capacity of the soil.
- nutrients–all nutrient elements required for the growth of higher plants are needed for microbial growth. Adequate nutrients are present in most fertile soils but additional nitrogen must be added if a highly carbonaceous waste (C/N ratio >35) is applied.

Degradation of industrial wastes by aerobic microorganisms is preferred. Aerobic degradation rates are faster than anaerobic processes. Also, the products of aerobic degradation are less odiferous in comparison to anaerobic decomposition. Wastes are kept aerated at soil incorporation sites by frequent mixing.

Nonbiological Degradation

Nonbiological degradation can be important in the dissipation of many organic substances in soils. In particular, evidence has been reported of the importance of chemical hydrolysis and photochemical degradation [5]. Oxidation-reduction is a major process for certain compounds.

Evaporation and Volatilization

Evaporation is a major mechanism of volume reduction at soil incorporation sites. The rate of evaporation from a wet soil is controlled by the same environmental conditions which control the rate of evaporation from standing water. Frequent mixing of the waste with soil increases evaporation and decreases the probability of developing anaerobic conditions.

Volatilization, although closely associated with evaporation, is the dissipation of waste constituents, rather than water, into the atmosphere. The magnitude of volatilization loss depends on the soil moisture content, chemical and physical properties of the waste and soil, atmospheric conditions (temperature, wind velocity, relative humidity, etc.), and application method. Mixing the waste with soil can reduce volatilization loss by increased adsorption of the chemicals to soil organic matter and clay, and the decreased vapor pressure of the waste.

VOLUMES AND TYPES OF INDUSTRIAL WASTE SUITABLE FOR SOIL INCORPORATION

The estimated volume of industrial sludges and wastewater that were suitable for disposal by soil incorporation in 1975 is shown in Figure 1. Also, projected volumes for 1980 and 1985 are indicated. These wastes represent approximately 3% of all industrial sludges and 1% of industrial wastewaters generated and projected for generation in the U.S. Thus, while soil incorporation could accommodate specific waste streams, the method will not significantly decrease industry's overall demand for other waste disposal methods.

Industries which produce wastes amenable for soil incorporation are listed in Table I. Industrial wastes are normally considered suitable for land cultivation if they comply with the following criteria:

1. the organic portion biologically decomposes at a reasonable rate;
2. does not contain material at concentrations toxic to soil microorganisms, plants or animals, and there must be reasonable assurance that long-term toxic effects resulting from accumulation through adsorption or ion exchange can either be prevented or mitigated;
3. does not contain substances in sufficient concentration to affect adversely the quality of the groundwater; and
4. does not contain substances in sufficient concentration to affect adversely soil structure, especially the soil infiltration, percolation and aeration characteristics.

The suitability of a specific industrial waste for land cultivation depends on many characteristics, including pH; biochemical oxygen demand (BOD); odor; concentrations of chemical elements (e.g., heavy metals and sodium), soluble salts and hazardous chemicals; bulk densities of waste solids; flammability; and volatility [6-8].

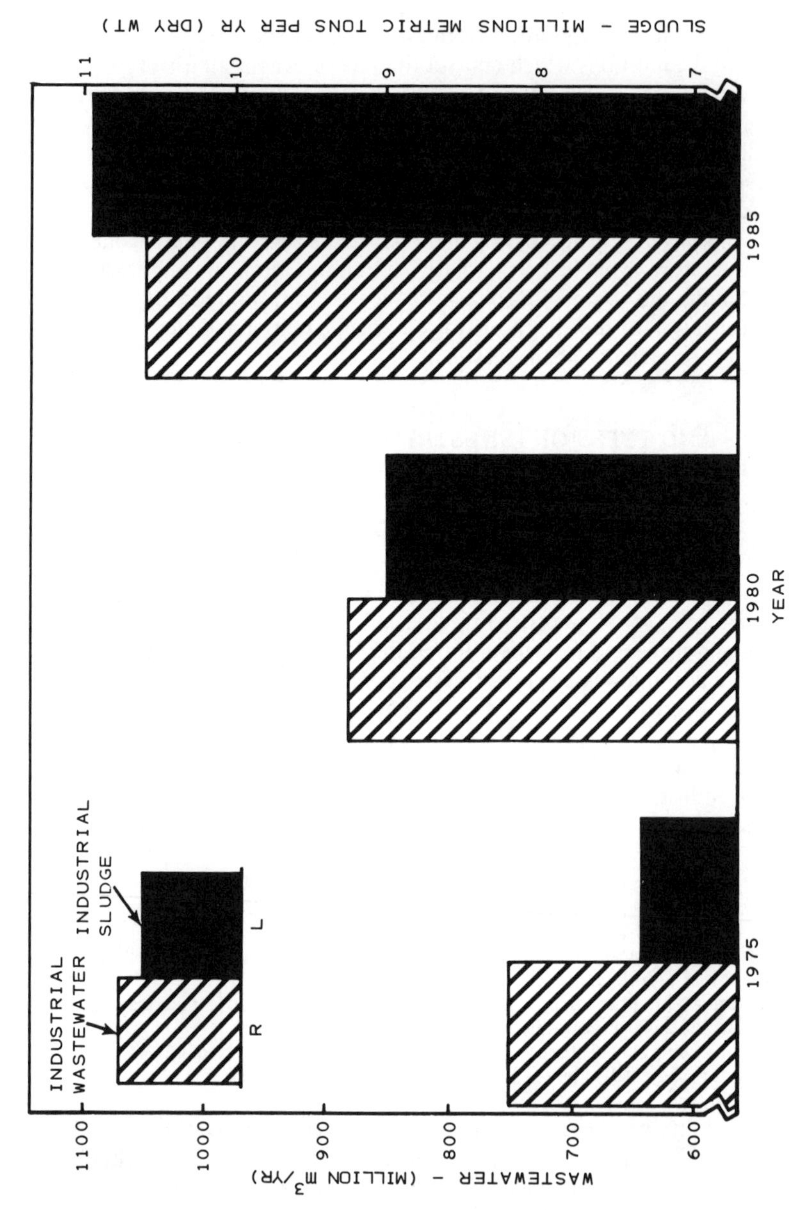

Figure 1. Volumes of industrial waste suitable for disposal by soil incorporation.

Table I. Industries That Produce Wastes Amenable to Soil Incorporation [2]

Food and kindred products
Textile finishing
Wood preserving
Paper and allied products
Organic fibers (noncellulosic)
Drugs and pharmaceuticals
Soap and detergents
Organic chemicals
Petroleum refining
Leather tanning and finishing

ENVIRONMENTAL IMPACTS

Data on environmental contamination from soil incorporation of industrial wastes are scarce, since the disposal method per se has not been monitored over an extended period. Much of the assessment of soil incorporation and its environmental impacts must therefore be based on assumptions and data from recent, short-term studies.

When waste is incorporated into a surface soil, a series of physical, chemical and biological processes is triggered. Phillips and Nathwani [9] reviewed the various mechanisms involved in soil-waste interactions. Also, SCS Engineers evaluated impacts at 11 case-study sites [2]. Of significance are the potential impacts on groundwater and surface water, air quality, soil, vegetation, and human and animal health.

Water Quality Impacts

Wastes applied to surface soils are susceptible to washout by precipitation runoff. Soil incorporation designs include appropriate drainage control facilities to prevent such occurrences, including upstream runoff diversion channels and settling basins to contain onsite any waste-laden runoff water. No evidence of surface water contamination due to runoff from a soil incorporation site is documented, although flooding has been reported. Flood problems are due to poor site-selection rather than inherent shortcomings of the disposal method itself.

As with any land disposal operation, groundwater quality may be impaired if surface-applied waste constituents are allowed to move through the subsurface soil to underlying aquifers. Soils exhibit tremendous chemical and physical attenuation and degradation capacity of the added waste due to their high reactive surface area and the varied microbial populations. However, if a waste has soluble or partly soluble constituents or if the soil is unable to retain certain constituents due to overloading, there is always a risk that it will affect the quality of subsurface waters.

The downward movement of heavy metals, oils and pesticides in soils is often restricted due to low water solubility, high retention and degradation by various soil processes [10-12]. Other chlorinated hydrocarbons, phenols and detergent components may be present in varying amounts in the wastes, depending on the generator industry. It is believed that such organic compounds will be eventually decomposed by soil microorganisms. Precise decomposition rates and favorable environmental conditions are not known. Unless the soil is overloaded with wastes containing large amounts of these substances, soil incorporation is not likely to pose a serious threat to groundwater quality.

Air Emissions

Certain industrial wastes can emit odors and volatile compounds when exposed to the atmosphere, impairing air quality in the disposal area and possibly endangering public health. If the waste contains volatile components, soil incorporation practices could increase the losses of these components during spreading. However, some of the readily volatile fraction likely would have come off during waste handling before delivery to the disposal site. Additionally, during remixing, dust could present a health hazard to the personnel on the site and may be carried offsite by winds. Subsurface injection of the waste and/or mixing with soil as soon as practical after deposition can alleviate, but not always eliminate, odors and volatilization. The proposed EPA Hazardous Waste Guidelines and Regulations would restrict the type and amount of volatile wastes that could be applied based on a prediction of the potential for air emissions [13].

Soil Impacts

Composite samples of surface soil were taken by SCS Engineers from several soil incorporation sites [2]. Samples were also collected from corresponding control sites which had similar soil and vegetation, but had received no waste. The objective of the limited sampling program was to determine the effect of soil incorporation on soil chemical properties. Accumulation of sludge-borne metals in the treated soils is shown in Table II. Addition of refinery sludge resulted in significant increases in the levels of 0.1 *N* HCl-extractable sulfate and metals (except for molybdenum) in the surface soil. Significant increases were noted for extractable orthophosphate, manganese, nickel and lead due to treatment of tannery sludge. Although the chromium in the tannery waste was high (2.25%), its concentrations in the treated surface soils were only 3.2 and 1.1 μg/g, respectively, from the clover- and cornfields. Apparently, the dilute acid-extractable chromium does not reflect total chromium in soil.

Table II. Chemical Analysis of Surface (0-15 cm) Soils of Control and Sludge-Treated Fields (μg/g) [2]

	Refinery Sludge[a]		Tannery Sludge[b]			
Element[c]	Control	Treated	Control	Treated	Control	Treated
Ortho-P	5.6	6.3	1.7	6.5	4.4	13.5
SO_4	18.7	625	22	7	2.5	2.5
Mo	0.27	0.24				
Fe	6.8	9.8	22.7	24.0	22.0	17.0
Mn	22.3	70	131	180	103	154
Cu	0.5	26.8	5.81	4.31	4.12	5.44
Ni	1.0	2.3	1.83	2.33	<0.07	1.00
Pb	1.2	160	3.0	6.0	4.83	6.67
Cr	0.08	11.3	1.0	3.2	0.6	1.1
Cd	0.03	0.11	0.08	0.17	0.10	0.12
Co	0.20	0.62	1.22	1.44	0.97	1.06
Ag	0.06	0.17				
Al	140	135				
Ti	0.35	2.90				
Zn	1.5	118	12.3	12.5	9.8	12.3

[a]Applied to a silt loam (pH 6.6) and planted to winter wheat.
[b]Applied to a slit loam (pH 6.3 and 5.9) and planted to clove (columns 3 and 4) and corn (columns 5 and 6).
[c]0.1 *N* HCl-extractable.

Vegetative Impacts

At ongoing soil incorporation sites receiving industrial wastes, existing vegetation is usually removed from the site before waste application. Weeds and small bushes will reestablish in the disposal plot only if the plot is left untilled for some time. Limited data are available from greenhouse and field investigations conducted to evaluate the potential adverse crop effects of applying certain industrial sludges on land. DeRoo [14] evaluated the use of mycelial sludges produced by the pharmaceutical industry in Connecticut as a nitrogen fertilizer and organic soil amendment. He concluded that if the mycelial sludge is applied repeatedly at high rates (100 ton/ac) to the same field, the soluble salt concentration and high zinc content in the sludge may be injurious to plants. Studies with similar objectives have been conducted using lagooned paper pulp sludge [15], cannery fruit sludge [16] and nylon sludge [17]. Results indicate that these sludges would have value as a low-analysis nitrogen fertilizer, and that no adverse effects were observed in crops and soils at normal application rates.

A preliminary analysis was conducted by SCS Engineers of composite plant samples from six soil incorporation sites receiving industrial wastes. As noted in Tables III and IV, plant analyses at two sites show that some increases in heavy metal uptake were noted, particularly lead (oil refinery) and chromium (tannery). The concentrations of trace elements were not at levels considered either phytotoxic or unsafe for animal consumption.

Table III. Trace Element Concentrations (μg/g) in Wheat Grain Grown on a Silt Loam Amended with Refinery Sludge [2]

Element	Control Plot[a]	Treated Plot			
		1	21	27	Avg.
B	1.38	2.75	2.75	2.12	2.54
No	1.69	2.45	4.33	3.39	3.39
Fe	48.3	55.0	48.3	50.4	51.2
Mn	21.67	37.50	41.25	40.00	39.58
Cu	3.81	5.38	5.75	5.38	5.50
Ni	2.5	5.0	5.0	2.75	4.25
Pb	<0.05	1.0	<0.05	1.23	
Cr	0.42	3.13	2.5	2.92	2.85
Cd	0.11	0.15	0.18	0.20	0.18
Co	0.19	0.06	0.15	0.06	0.09
Ag	0.33	0.33	0.29	0.13	0.25
Al	0.68	0.50	1.40	1.31	1.07
Ti	0.73	0.26	0.73	0.38	0.46
Zn	61.0	72.0	73.8	70.0	72.2

[a]Composite sample was made from three control plots.

Table IV. Trace Element Concentrations (μg/g) in Clover and Corn Grown on a Silt Loam Amended with Tannery Sludge [2]

Element	Clover		Corn Kernel	
	Control	Treated[a]	Control	Treated[a]
B	7.88	5.7		
Se	<0.016	0.05		
Mo	5.06	2.38		
Fe	712	464	97.3	81.8
Mn	120	119	6.43	7.04
Cu	12.27	12.24	2.23	1.86
Ni	28.75	10.52	8.60	8.98
Pb	3.12	7.18	2.56	2.34
Cr	32.2	77.1	5.9	4.9
Cd	0.09	0.20	0.04	0.04
Co	1.56	0.52	0.33	0.30
Zn	63.75	72.82	32.79	29.88

[a]Average of three composite samples.

PROPOSED REGULATIONS

EPA issued proposed guidelines and regulations on December 18, 1978 [13], for hazardous waste identification, transportation and disposal. These regulations cover various disposal methods including land farms, i.e., sites where soil incorporation is practiced (Paragraph 250.45-5). Table V summarizes the regulations for land farms.

Although these proposed regulations are subject to revision, the basic control philosophy will likely remain unchanged. It appears that the law will allow disposal of many types of industrial wastes, including hazardous waste, by soil incorporation if the appropriate conditions are met. Of course, individual states can impose more stringent controls on this and other waste disposal practices.

IMPLEMENTING A SOIL INCORPORATION SITE

Site-Selection

Sites used for soil incorporation must conform with all requirements imposed on land used for other disposal methods. EPA has issued proposed criteria for determining whether a site is suitable for land disposal [18]. These sanitary landfill criteria (also applicable to soil incorporation sites) are highlighted below:

- Environmentally sensitive areas: a disposal site shall not be located in the following areas: (1) wetland; (2) floodplain; (3) permafrost; (4) critical animal habitats; or (5) over sole-source aquifers
- Ground- and surface waters cannot be endangered
- Air emissions must comply with all applicable regulations and must not adversely affect public health
- Crops cannot be grown on lands treated with wastes that contain excessive concentrations or amounts of cadmium, pathogens, pesticides or other organics. Amounts for cadmium are specified
- Disease vectors must be controlled
- Public safety must be ensured: (1) toxic or explosive gases must be controlled; (2) fires must be quickly extinguished; and (3) access must be controlled

Some of the above-noted conditions must be met by proper site location; other conditions can be satisfied by appropriate design features.

From a practical standpoint, a soil incorporation site must be:

1. readily accessible to all intended users;
2. compatible with surrounding land use;
3. physically suitable for the intended operations, i.e., the topography, hydrogeology, climate and soils must be appropriate;
4. cost-effective to initiate and operate; and
5. politically and socially acceptable to the community.

Table VI indicates the range of site soil conditions that could be encountered and the degree to which these conditions could affect soil incorporation operations. Local climatic conditions can significantly

Table V. Summary of Proposed EPA Hazardous Waste Regulations Pertaining to Soil Incorporation (Landfarming) [13]

Hazardous wastes not amenable to landfarming[a]

- Ignitable waste
- Reactive waste
- Volatile waste
- Wastes which are incompatible when mixed

General requirements

- Wastes shall not contact navigable water
- The treated area shall be at least 1.5 m (5 ft) above the historical high groundwater table[b]
- Erosion shall be minimized
- The treated area shall be at least 150 m (500 ft) from water supplies[b]
- Onsite soils shall be fine-grained, within specified limits[b]

Site preparation

- Surface grades shall be less than 5%[b]
- Direct connections (such as caves and wells) on the site and within 30 m (100 ft) thereof shall be sealed[b]
- Soil pH shall be greater than 6.5

Waste application

- Waste incorporation practices shall maintain aerobic conditions in the surface soils
- Waste shall not be applied to a saturated soil
- Waste shall not be applied when soil temperature is 0°C or below
- The soil-waste mixture shall be maintained at pH 6.5 or greater until site closure[b]
- Any nitrogen and phosphorus added to the soil shall not exceed U.S. Department of Agriculture application rates

Soil monitoring

- Background soil conditions shall be determined
- Soil conditions in the treated area shall be monitored semiannually by sampling one core per acre
- If soil monitoring indicates downward movement of contaminants, regulatory officials shall be notified and landfarming ceased

Crop growth

- Food chain crops shall not be grown on the treated area of a hazardous waste land farm

Closure

- A completed site shall be returned to preexisting conditions
- If the soil is itself a hazardous waste, it shall be removed for management as such[b]

[a]If air emissions of contaminants are within given limits, certain such hazardous wastes could be land farmed.

[b]This requirement can be modified if specific site, design or operational conditions indicate adequate environmental protection is still attainable.

Table VI. Soil Limitations for Accepting Nontoxic Biodegradable Industrial Wastes [19]

Item	Degree of Soil Limitations		
	Slight	Moderate	Severe
Permeability of the most restricting layer above 152 cm	Moderately rapid and moderate (1.5-15 cm/hr)	Rapid and moderately slow (15-51 and 0.5-1.5 cm/hr)	Very rapid, and very slow (>51 and <0.5 cm/hr)
Soil drainage	Well drained and moderately well drained	Somewhat excessively drained and somewhat poorly drained	Excessively drained, poorly drained and very poorly drained
Runoff	None, very slow and slow	Medium	Rapid and very rapid
Flooding	Soil not flooded during any part of the year		Soil flooded during some part of the year
Available water capacity from 0 to 152 cm or to a limiting layer	>20	7.6-20 cm	<7.6 cm

[a]For definitions see the *Soil Survey Manual*, USDA No. 18 (1951).

influence the viability of this disposal practice. Excessively high moisture content in the soil may impede oxygen transfer to soil microorganisms, thereby slowing waste degradation. Also, from a practical standpoint it is difficult to properly mix waste into a muddy soil. Soil and climatic conditions are but two important site-selection parameters to be evaluated.

Site Operational Considerations

Equipment and Personnel Needs

The quantity and types of equipment and personnel required depends on the volume and characteristics of waste to be disposed, requirements imposed by regulatory agencies, land area of the site and need for other duties (e.g., to direct traffic and unloading operations).

Equipment selection depends primarily on the characteristics of the waste applied and on regulatory constraints. Tank wagons may be adequate for surface application of sludge at some locations, whereas subsurface injection equipment may be required at other locations by local regulations. Disk plows are suited for mixing liquid waste and sludges. In some cases waste can be effectively mixed by running a track dozer over the soil; the blade and tracks turn over the soil and help aerate the mass. Other agricultural equipment may be satisfactory to some extent, especially for sites that handle small quantities of waste.

Waste Storage

Field operations are usually curtailed during periods when the soil is excessively wet or frozen. Wastes are stored for application under more favorable conditions. Storage is also necessary when an equipment breakdown occurs. Storage facilities can be provided at the treatment site itself or at the generator's location.

Waste Loading and Reapplication

Waste composition and percent solids generally dictate the amount of waste that can be applied to a soil incorporation site. The waste loading rate can be determined according to the partial requirement of crops for nitrogen and/or phosphorus such as for paper mill and fruit cannery sludges. High concentrations of such constituents as sodium, soluble salts, sulfide and heavy metals may limit the quantity of waste that can be safely disposed of on land.

If a soil cannot assimilate the applied waste, it will become anaerobic, resulting in nuisance conditions and possible failure of the system to effectively degrade the organic matter. Furthermore, unless the wastes are detoxified or decomposed by the soil or weather to nondeleterious products, repeated applications will eventually load the upper soil zone to its ultimate capacity. As a result, disposal activities at the site will have to be terminated and the site may be unusable indefinitely for alternative purposes.

Loadings of industrial wastes can vary considerably. Wallace [20] summarized the hydraulic and organic loading rates used at existing land application sites for industrial wastewaters (Table VII). He noted that either hydraulic or organic loading can control in a given instance. In some cases, neither will be as important as the cation loading, especially with regard to sodium. For wastewaters with biological oxygen demand (BOD) concentrations in the 1000-mg/l or less range, hydraulic loading will usually control.

Loading rates of industrial sludge depend on the waste composition, site characteristics and crops to be grown, if cropping is planned. Except for oil refinery sludges, data on application rates for industrial sludges are meager.

Refinery sludge is usually spread to a layer 7.6-15.2 cm thick. Lewis [11] reported sludge at an average of 1008 metric ton/ha/yr (450 ton/ac/yr) was

Table VII. Summary of Hydraulic and Organic Loading Rates Used in Existing Land Application Systems for Industrial Wastes [20]

	Hydraulic Load		Organic Load (BOD)
Type of Waste	(gal/ac/day)[a]	(in./wk)	(lb/ac/day)[b]
Biological Chemicals	1,500	0.39	370
Fermentation Beers	1,350	0.35	170
Vegetable Tanning			
Summer	54,000	13.91	360
Winter	8,100	2.09	54
Wood Distillation	6,850	1.76	310
Nylon	1,700	0.44	287
Yeast Water	15,100	3.89	–
Insulation Board	14,800	3.81	138
Hardboard	6,000	1.55	85
Boardmill Whitewater	15,100	3.89	38
Kraft Mill Effluent	14,000	3.61	26
Semichemical Effluent	72,000	18.54	90-210
Paperboard	7,600	1.96	13-30
Deinking	32,400	8.34	108
Poultry	40,000	10.30	100
Peas and Corn 57-Day Pack	49,000	12.62	238
Dairy			
Low Value	2,500	0.64	10
High Value	30,000	7.73	1,000
Soup	6,750	1.74	48
Steam Peel Potato	19,000	4.89	80
Instant Coffee and Tea	5,800	1.49	92
Citrus	3,100	0.80	51-346
Cooling Water–Aluminum Casting	95,000	24.46	35

[a]Multiply by 9.35×10^{-3} to convert to m^3/ha/day.
[b]Multiply by 0.89 to convert to kg/ha/day.

applied at one refinery. This amounts to 112 metric ton/ha/mo (50 ton/ac/mo) based on a 9-mo operation. Generally, when the oil content in the surface 15 cm (6 in.) decreases to 2-4%, additional sludge can be applied. Under normal operating conditions this would take about 2 mo. Waste application on such a frequency precludes growth of cover vegetation. Optimum application rates for oily waste appear to be about 5% oil in soil [21].

Soil Amendments

It is necessary to ensure that adequate nutrients are present in the soil for optimum microbial decomposition of wastes. Most nutrients (except, probably, nitrogen) are usually abundant in soils. Periodic additions of nutrients may be necessary to maintain an acceptable degradation rate, especially when highly carbonaceous wastes are applied. At many soil incorporation sites, operations are conducted without application of any nutrients.

Lime may be needed to adjust soil and waste pH to near neutral (pH 7). Movement of most heavy metals and phosphorus away from the incorporation zone is minimized at the higher pH levels. Liming acid soils also favors microbial activity.

Mixing the Waste with Soil

Most industrial sludges are spread or leveled to a thin layer soon after application and allowed to dry for a few days. After initial drying, the material is incorporated into the soil by one or more diskings. Depending upon the visual appearance of the decomposition products and weather conditions, additional disking and drying periods are completed as necessary.

If wastewater or dilute liquid sludge is applied to a soil, especially when a cover vegetation is present, disking may be unnecessary. Subsurface injection of industrial sludge can result in partial mixing of the waste with soil, and generally disking is not required.

At most soil incorporation sites receiving oily sludges, initial drying generally takes 1-3 wk depending on waste loading and evaporation rates. The waste is then mixed into the soil by a track dozer and/or disking. Subsequent mixing is done at weekly to monthly intervals. The waste-treated soil is sometimes disked and leveled as part of site preparation prior to receiving additional waste.

Site Management

Management of a soil incorporation site involves several factors.

Soil management. The goals of soil management are to immobilize heavy metals and other toxic constituents, increase the decomposition of waste materials, control erosion and runoff, and prevent groundwater contamination. The goals can be attained by maintenance of (1) neutral to alkaline conditions; (2) adequate nutrient concentrations; and (3) surface slopes and runoff control facilities.

Timing of operations. Soil incorporation scheduling must account for wind, precipitation and air temperature. For instance, when the direction of the wind can cause odors that may affect inhabited areas, operations should be suspended. Waste storage capacity is required for when freezing temperatures do not permit winter operation, or where application on snow-covered ground is impractical. If the site is used for crop cultivation, schedules for water application must be adjusted according to rainfall to avoid overirrigation.

Other management considerations. These involve control of wind movement and snow distribution (by screening or fencing), and site monitoring.

Monitoring

A monitoring program for a soil incorporation facility is best tailored to specific waste and site conditions. In general, water and air quality protection are of primary concern. If volatile wastes are to be handled, air emissions should be monitored. Any crops for human or animal consumption grown on waste-treated soil should be analyzed periodically to ensure that potentially harmful levels of contaminants do not accumulate in edible portions of plant tissue.

Monitoring for groundwater quality control may be simpler and less costly at a soil incorporation site when compared to a sanitary landfill. Since wastes are applied only to the surface soil, routine analysis of surface and near-surface soil samples can indicate if and when contaminants move away from the near-surface zone of waste incorporation. Any movement of contaminants can be detected and remedial action taken well before underlying groundwater has been contaminated. Routine groundwater monitoring by wells is generally not necessary except to establish background water quality data before site activation. Soil monitoring is the indicated monitoring method for landfarms in the EPA's proposed hazardous waste regulations [13].

Surface water can be monitored at the siltation basin. At a properly constructed and maintained site the siltation basin will accumulate any surface water that occurs at a soil incorporation facility.

ECONOMICS

The unit costs of industrial waste disposal for five case-study soil incorporation sites are shown in Figure 2. Also shown is the expected range of costs for a hypothetical site, estimated on the basis of a conceptual design prepared by SCS Engineers [2]. Unit costs for both actual case studies and the conceptual design include expenditures for land, equipment, labor, interim onsite storage of waste and environmental monitoring. Cost for waste transport to the site is not included.

Capital costs for the conceptual design range from $262,000 to $746,000 for waste delivery rates of 5340-21,370 gal (1000-4000 dry metric tons) per

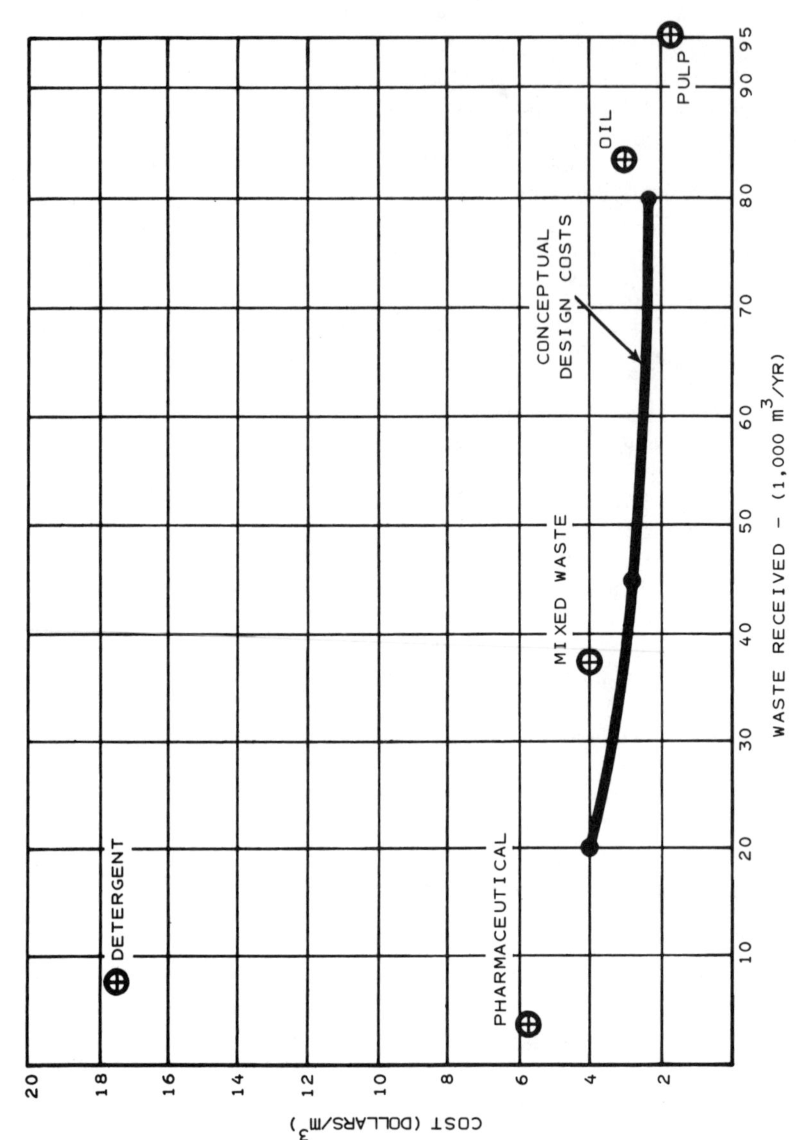

Figure 2. Costs of hypothetical conceptual design for a land cultivation site compared to case study costs.

year (for sludge at 5% solids). Unit costs range from \$0.01 to \$0.07/gal (\$2.00 to \$18/m^3) for soil incorporation.

By comparison costs for disposal of uncontained industrial wastes by sanitary landfill methods generally range from \$0.02 to \$0.08/gal (\$5.30 to \$22.00/m^3), depending on location, site volume and other factors. For industrial waste disposed of in drums or other containers, costs can be substantially higher, ranging up to \$0.72/gal (\$194/m^3). Thus, soil incorporation cost can be comparable with or lower than conventional disposal methods.

SUMMARY

Industrial wastes that have been disposed of by soil incorporation are primarily from food processing, oil refinery, paper and pulp, tannery and pharmaceutical industries. The wastes are composed mainly of organic material and are thus biodegradable. The soil incorporation practice is limited and will likely remain limited in applicability to about 3% of all industrial wastes. Soil incorporation is viable only where soil, geology, waste characteristics, climate and other environmental conditions permit. Depending on specific conditions, the disposal program can either be related to agriculture or can be solely a disposal practice. Proposed EPA regulations would control, but not necessarily eliminate, disposal of hazardous wastes by soil incorporation.

Documented environmental impacts associated with soil incorporation operations concern overloading soil with waste constituents, plant uptake of heavy metals and emanation of odors. These and other potential impacts may be controllable by improved operating techniques and effective monitoring programs. In particular, routine soil monitoring to detect any movement of waste constituents away from the zone of incorporation can provide an early warning of fugitive contamination. In the face of a substantial threat to groundwater, all waste and surface soil at a soil incorporation site could be stripped and moved to other sites.

Soil incorporation costs are in the range of \$0.01-0.07/gal (\$2-18/m^3) of industrial waste, varying with waste type, loading rates, site use, site area and other methods of industrial waste disposal.

REFERENCES

1. Powers, P. W. *How to Dispose of Toxic Substances and Industrial Wastes* (Park Ridge, NJ: Noyes Data Corporation, 1976).
2. Phung, T., L. Barker, D. Ross and D. Bauer. "Land Cultivation of Industrial Wastes and Municipal Solid Wastes: State-of-the-Art Study–Volume I," EPA-600/2-78-140a, U.S. EPA (1978).
3. Alexander, M. *Introduction to Soil Microbiology*, 2nd ed. (New York: John Wiley and Sons, Inc., 1977).
4. Martin, J. P., and D. D. Focht. "Biological Properties of Soils," in *Soils for Management of Organic Wastes and Waste Waters.* (Madison, WI: American Society of Agronomy, 1977), pp. 115-169.

5. Armstrong, D. E., and J. G. Konrad. "Nonbiological Degradation of Pesticides," in *Pesticides in Soil and Water*, W. D. Guzeni, Ed. (Madison, WI: Soil Science Society of America, Inc., 1974), pp. 123-131.
6. Epstein, E., and R. L. Chaney. "Land Disposal of Industrial Waste," in Proceedings of the National Conference on Management and Disposal of Residues from the Treatment of Industrial Wastewaters, Washington, DC, February 1975, pp. 241-246.
7. Wallace, A. T. "Land Disposal of Liquid Industrial Wastes," in *Land Treatment and Disposal of Municipal and Industrial Wastewater*. R. L. Sanks and T. Asano, Eds. (Ann Arbor, MI: Ann Arbor Science Publishers, Inc., 1976), pp. 147-162.
8. Lehman, M. P. "Industrial Waste Disposal Overview," in Proceedings of the National Conferenence on Management and Disposal of Residue from the Treatment of Industrial Wastewaters, Washington, DC, February 1975, pp. 7-12.
9. Phillips, C. R., and J. Nathwani. "Soil-Waste Interactions: A State-of-the-Art Review," Solid Waste Management Report, EPS 3-EC-76-14, Environmental Canada (1976).
10. Letey, J., and W. J. Farmer. "Movement of Pesticides in Soils," in *Pesticides in Soil and Water*, W. D. Guenzi, Ed. (Madison, WI: Soil Science Society of America, Inc., 1974), pp. 67-97.
11. Lewis, R. S. "Sludge Farming of Refinery Wastes as Practiced at Exxon's Bayway Refinery and Chemical Plant," in Proceedings of the National Conference on Disposal of Residues on Land, St. Louis, Missouri, September 1976, pp. 87-92.
12. Page, A. L. "Fate and Effects of Trace Elements in Sewage Sludge When Applied to Agricultural Lands," EPA-670/2-74-005, U.S. EPA (1974).
13. "Hazardous Waste Proposed Guidelines and Regulations and Proposal on Identification and Listing," *Federal Register*, 43(243) (December 18, 1978).
14. DeRoo, H. C. "Agricultural and Horticultural Utilization of Fermentation Residues," Connecticut Agricultural Experiment Station, Bulletin No. 750 (1975).
15. Jacobs, L., Michigan State University, East Lansing, MI. Personal communication.
16. Noodharmcho, A., and W. J. Flocker. "Marginal Land as an Acceptor for Cannery Waste," *J. Am. Soc. Hort. Sci.* 100:682-684 (1975).
17. Cotnoir, L., University of Delaware, Neward, DE. Personal communication.
18. "Solid Waste Disposal Facilities, Proposed Classification Criteria," *Federal Register*, 43(25) (February 6, 1978).
19. Soil Conservation Service, U.S.D.A. "Guide for Rating Limitations of Soils for Disposal of Wastes," Interim Guide, Advisory Soils–14. Washington, DC (1975).
20. Wallace, A. T. "Land Disposal of Liquid Industrial Wastes," in *Land Treatment and Disposal of Municipal and Industrial Wastewater*, R. L. Sanks and T. Asano, Eds. (Ann Arbor, MI: Ann Arbor Science Publishers, Inc., 1976), pp. 147-162.
21. Phung, H. T., and D. E. Ross. "Soil Incorporation of Petroleum Wastes," *Water 1978* (1979).

INDEX